Night Owls and Early Birds

Philippa Gander

Night Owls and Early Birds

Rhythms of Life on a Rotating Planet

AUCKLAND
UNIVERSITY
PRESS

First published 2023
Auckland University Press
University of Auckland
Private Bag 92019
Auckland 1142
New Zealand
www.aucklanduniversitypress.co.nz

ISBN 978 1 86940 978 4

A catalogue record for this book is available from the
National Library of New Zealand

Design by Greg Simpson
Illustration input by Tim Nolan / Blackant Mapping Solutions

This book was printed on FSC® certified paper

Printed in China through Asia Pacific Offset Group Ltd

Contents

Preface

I have had the great good fortune to work for more than 40 years in a job that I am still passionate about. When people ask what I do for a living, I say I am a research scientist. Typically, the next question is 'what do you study?'. If I say chronobiology, the next question is almost invariably some (more or less polite) version of 'what on Earth is that?'. This book seeks to answer that question and explain my passion for chronobiology, which I see as critical to our understanding of all life on Earth, including our own health, safety, and well-being every day of our lives.

The book is a personal view across a broad vista of science. Many great colleagues have accompanied various parts of my serendipitous journey as a chronobiologist and could write aspects of this narrative with far greater authority than I can. Their collegiality, insight, and hard work continue to fuel my passion for trying to understand the innate rhythmicity of life forms on our rotating planet, with its orbiting moon and its annual orbit around the sun.

I would never have begun this journey if I had not enrolled, as a somewhat distracted second-year science student, in Bob Lewis's zoology paper 'Animal Orientation and Timing'. The initial attraction was that it fitted into my timetable around the courses I had to study to major in molecular biology. Bob turned out to be an inspirational teacher, I switched to majoring in zoology, and he became my postgraduate supervisor and mentor.

I am indebted to many great colleagues from science, industry, and government agencies with whom I have worked over the years and who taught me so much. I have also learned a great deal and had many wonderful times with the colleagues and students who grew with me through the last 20-plus years of research at the Sleep/Wake Research Centre.

I gratefully acknowledge the friends and family who gave valuable feedback on earlier drafts: Anneke Borren, Michael Cuncannon, Curt and Jane Graeber, and Adrian Moonen.

I am indebted to Sam Elworthy at AUP for his vision for the book and for seeing it through in complex times. Katharina Bauer was a pleasure to work with as production editor, as was Matt Turner who provided thoughtful editing that enhanced the text. Many thanks to Louise Belcher and Sophia Broom for proofreading, to Tim Nolan for an outstanding job converting the original colour figures to black and white, and to Greg Simpson for the book design.

This book is dedicated to my children and their partners: Perrine Boy and Simon Bell, Jeremy Boy and Pauline Gourlet, and their children Leon and Max Bell, and Noam Boy. It is also dedicated to our beautiful planet Earth that has cradled life as we know it and has so much more to teach us if, with humility, we can open our hearts as well as our minds.

Introduction

This book is about time – biological time. It is about the innate rhythms of life on planet Earth, which at the time of this writing is still the only planet that we know for certain has life on it.

The physical environment on Earth is not constant – it fluctuates regularly with the geophysical cycles. Earth's rotation on its axis gives us the 24-hour day/night cycle. The moon's orbit around Earth gives us the monthly lunar cycle and the tides. Earth's yearly orbit around the sun gives us the cycle of the seasons. Life here is adapted to the timing of our planet, just as it is adapted to Earth's atmosphere and gravity. Living organisms have internal timekeeping systems – endogenous biological clocks – to take advantage of the opportunities, and to manage the challenges, that come with predictable changes in the physical environment. These internal timekeeping systems are the focus of the scientific discipline of chronobiology – literally the biology of time.

In 1975, when I decided to do my postgraduate research training in chronobiology, one of our physiology professors announced that it was a lunatic fringe discipline and a waste of a good student. Fast-forward 42 years, and in 2017 the chronobiologists Jeffrey Hall,

Michael Rosbash, and Michael Young were awarded the Nobel Prize for Physiology or Medicine 'for their discoveries of molecular mechanisms controlling the circadian rhythm'.

Circadian rhythms (from the Latin *circa* – about, *dies* – day) are an adaptation to Earth's daily rotation and they occur in all cell-based life forms on our planet. We share this adaptation with ancient single-celled organisms like cyanobacteria (blue-green algae), which have DNA but no cell nucleus and have been around for more than two billion years. The core molecular mechanisms behind circadian rhythms are known as circadian clocks. They consist of a small number of clock genes that interact to produce circadian rhythms in concentrations of clock proteins, with a cycle length of about 24 hours. In turn, the circadian rhythms in clock proteins drive circadian rhythms in a multitude of clock-controlled genes, resulting in cascades of circadian rhythms in the functioning of cells throughout the body.

The molecular workings of circadian clocks are currently better understood than those of other biological clocks. There is evidence in some organisms for internally generated biological clocks that are adaptations to the complex environmental cycles generated by the moon, including circatidal clocks (adaptations to the ocean tides) and circalunar clocks (adaptations to the cycles of the moon). Some species also have internally generated circannual clocks (adaptations to the seasonal changes caused by Earth's yearly orbit around the sun).

The circadian clock genes are present in every cell in the human body and are active in most of them. We literally have millions of cellular circadian clocks, whose activity is coordinated by a hierarchical system that produces daily rhythms in our physiology and behaviour. This circadian timekeeping system may also contribute to the regulation of our monthly and seasonal rhythms, such as the bi-monthly mood changes experienced by some people with bipolar disorder, seasonal depression, and fluctuations in birth rates.

Chronobiology has come a long way. So has sleep science, highlighting the vital importance of this third of the daily cycle. However, the idea that living organisms are intrinsically rhythmic, and that sleep is an integral part of every aspect of our health, safety, and well-being, is still a long way from being accepted in mainstream thinking by science, medicine, or society.

We are relative late-comers in the story of life on Earth and we mistakenly think that our ability to modify the environment and, more recently, our use of technology have enabled us to separate ourselves from its cycles.

This book examines evidence that we may be paying a high price for overlooking the innate rhythmicity of life on Earth, and it refutes the widespread myth that sleep is an 'off time' that can be sacrificed for getting more out of a busy waking life. For example, shift work is the most common major disruption to our circadian rhythms, affecting more than 20 per cent of the global working population. Large studies that have tracked the health of night-shift workers over time indicate that they are at greater risk than their non-shift working colleagues of developing obesity, type 2 diabetes, high blood pressure, cardiovascular disease, some types of cancer, dementia, and of dying younger from all causes.

The evidence is now clear that getting enough sleep on a regular basis is one of the three pillars of health, along with diet and exercise. Indeed, the legal argument is now being made that healthy sleep is a fundamental human right under the United Nations Universal Declaration of Human Rights. In the context of the Covid-19 epidemic, it is useful to know that the sleep you get on the night after vaccination is critical to the amount of immunity that you develop because the immune system recharges during sleep.

For us and many other species, the changes in light intensity that accompany the day/night cycle provide the key environmental time cue that keeps our circadian timekeeping system in step with Earth's rotation. Until very recently in the history of life on Earth, the day/night cycle, as well as the phases of the moon and the seasonal changes in daylight, provided reliable time cues to synchronise the internal rhythms of all cell-based life forms. Our extensive and rapidly expanding use of artificial light at night is being shown to disrupt entire ecosystems both on land and in the oceans. Our genetic programming for the geophysical cycles on Earth also creates interesting challenges for our dreams of living on the moon or Mars.

This book is an introduction to life on Earth viewed through the lens of chronobiology. It begins with Earth's geophysical cycles and the amazing diversity of internal timekeeping systems that have evolved in response to them, with examples from algae to humans. The focus then shifts to the circadian rhythms that modulate how we feel and function every day of our lives – how they are generated and how all our internal rhythms are synchronised with each other and kept in step with the environmental changes caused by Earth's rotation. Next, we explore the essential third of our lives that most of us know least about, namely sleep. Knowledge derived using modern scientific methods converges with old knowledge derived from careful observation and introspection. These chapters lay the foundation for an examination of the challenges that aspects of modern living generate for our health, safety, and well-being, as well as for other living systems.

Much of our modern understanding of life on Earth assumes that biological time is basically linear. Individual organisms progress from a beginning (splitting of a cell into two or combining the genes of two cells from different individuals), through development to reproductive maturity, a reproductive phase, and then degeneration and death. In this view, life is effectively a straight competitive trajectory from birth to death, driven by genetics interacting with natural and human selection processes. Chronobiology superimposes the idea that throughout every stage of this trajectory, all aspects of life fluctuate regularly in step with the geophysical cycles. Biological time is not linear, and life is more complex than we thought. However, the new understanding that comes from chronobiology also opens up new opportunities for improving people's lives and for reducing our negative impact on the complex, dynamic ecosystems that sustain us.

Welcome to the journey of this book. It explores both the challenges and some possible solutions generated by discoveries in the science of chronobiology, as well as some other types of knowledge that are essential for understanding the human condition and reducing our impact on our precious home planet.

Chapter One

Living in Earth's Geophysical Cycles

I grew up in the historic tourist town of Rotorua. It wasn't uncommon for jet-lagged 'foreigners' to have motor vehicle crashes because they were driving on the wrong side of the road (we drive on the left). My father, a policeman, was a shift worker. My mother was the second of 13 children and gardening, by necessity, was important for our extended family.

Looking back on it now, I see that my childhood gave me practical awareness of the challenges of flying across time zones and trying to work out of step with the day/night cycle. I can still visualise the seasonal changes in Mum's flowers and the family vegetable gardens. I recall the bitterly cold walk to school on some winter mornings and the summer joy of swimming in clean lakes. Perhaps this explains why, when I began studying chronobiology, it immediately seemed to make intuitive sense.

This chapter steps through some of the basic principles in chronobiology: how the day/night cycle, the lunar cycles, and the seasonal cycles arise, and some of the amazing adaptations that different species have developed to cope with these predictable changes in their environments.

A note on terminology. Throughout this book, the word 'cycles' is used to describe the regular fluctuations in the external environment (day/night cycles, tide cycles, lunar cycles, seasonal cycles). The word 'rhythm' always refers to an internally generated oscillation in some function of a living organism, which is an adaptation to a geophysical cycle. These internally generated rhythms typically do not have exactly the same periodicity as the geophysical cycles, so they have the prefix 'circa' (about). Circadian rhythms are about 24 hours; circatidal rhythms, about 12.4 hours; circalunar rhythms, about 28 days; and circannual rhythms, about a year. Time cues from the geophysical cycles lengthen or shorten these internal rhythms to keep them in step with the environmental cycles caused by Earth's rotation on its axis, its orbit around the sun, and the moon's orbit around Earth.

Daily (circadian) rhythms: legacy of Earth's rotation on its axis

As we go about our daily lives, the geophysical cycle we are most aware of is the 24-hour day/night cycle that results from Earth spinning on its axis. Daylight occurs while your part of Earth is facing towards the sun, which is about 150 million kilometres away, so sunlight takes about eight minutes to reach Earth. If you are at the equator, you are spinning at about 1667 kilometres per hour – fortunately we are not aware of this. The rotation speed decreases at higher latitudes, diminishing to zero at the poles. In fact, Earth's axis is slightly tilted with respect to the plane of its orbit around the sun. This tilt is what causes the seasons, as described later in this chapter.

Across the day/night cycle, there are marked changes in light, temperature, and humidity. The most favourable time of day for a given species to be active is influenced not only by the physical

Figure 1.1 The day/night cycle caused by Earth's rotation on its axis (not to scale) with the seasonal cycle caused by the axial tilt.

environment, but also by the rhythmicity of other species that are key to its survival, such as when its preferred food sources are available or when predators are active. Some species are nocturnal, some are diurnal, and some are active around dawn and dusk (crepuscular).

This raises a fundamental question at the heart of chronobiology: are these preferred times for activity simply a direct response to the day/night changes in the environment, or is there an internal timekeeping system that drives these preferences and is sensitive to the environmental changes? A key advantage of having an internal circadian clock is that it enables organisms to predict and prepare for regular daily changes in the physical environment and in important biological factors. As noted above, the main environmental time cue that synchronises the internal circadian clocks of most species is the changing light intensity across the day/night cycle.

One part of my PhD thesis research was designed to examine what controls the daily activity patterns of the Polynesian rat. This small rat is widely distributed throughout the Pacific and is thought to have been carried (intentionally or unintentionally) by the extraordinary Polynesian navigators who dispersed and settled in the different island groups. New Zealand is the southernmost landmass where Polynesian rats are found and are commonly known by their Māori name, kiore. During my research at the University of Auckland, I regularly spent time on beautiful Tititiri Matangi

Figure 1.2 An adult female kiore sitting on my hand. Photo: C. R. Austin

Island off the north-east coast, setting live traps at night to gather my nocturnal research participants. (On New Zealand's main islands, kiore did not manage to compete successfully with the larger Norwegian and ship rats that arrived later with European settlers.)

To see whether the nocturnal activity patterns of these kiore were driven by an internal circadian clock, we kept females in a controlled laboratory environment where all the time cues associated with the day/night cycle were carefully excluded. Figure 1.3 shows the activity patterns of one animal living in her own cage in the laboratory for a total of 87 days at a constant temperature of 21 degrees Celsius. When she moved through a dim red-light beam that crossed the cage, a pen resting on a scrolling paper strip was deflected. When she was moving around a lot, the pen moved back and forth frequently, producing the black 'activity' bars in the figure. Each horizontal line is a strip of paper that represents one day of recording. These pasted strips of paper from more than four decades ago seem a bit visually cluttered compared to the clean digital presentations we are now accustomed to, but they still tell a story.

Being nocturnal, in dim light this kiore was able to see well enough to move around, feed, and groom when she was awake. Across the 45 days in constant dim light (left panel of the figure),

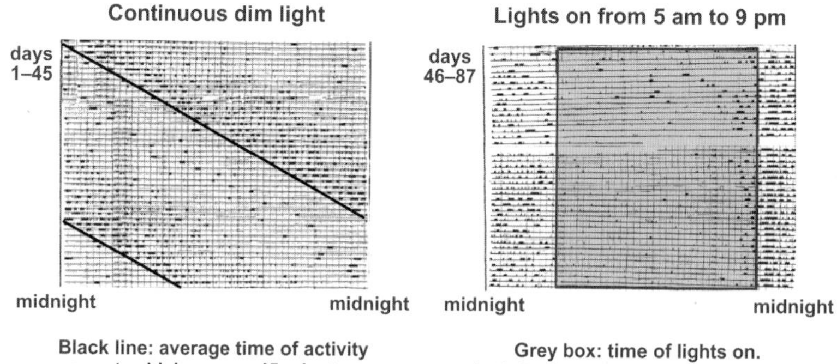

Figure 1.3 Activity rhythm of a female kiore living in a controlled laboratory environment.

she started becoming more active about 45 minutes later each day. In other words, the innate rhythm of her circadian clock in constant dim light was 24 hours and 45 minutes. This is shown by the black activity bars moving progressively to the right (each horizontal strip in the figure represents 24 hours).

From day 46 to day 87, bright lights were turned on from 5 am to 9 pm every day and the rat changed her pattern of activity. Every day she started becoming more active just before the lights went out and then reduced her activity just before the lights came on again. In the right panel of the figure, the black activity bars line up vertically down the page. Her circadian clock was synchronised to exactly 24 hours by the artificial 24-hour light/dark cycle.

Here we see two essential features of circadian clocks: 1) without any 24-hour time cues from the environment, the innate rhythms of circadian clocks are typically not exactly 24 hours; but 2) they can be synchronised to exactly 24 hours by a 24-hour light/dark cycle.

The circadian clocks of kiore control many other aspects of their lives in addition to their daily patterns of activity. The estrous cycle of female rats is regulated by a daily hormonal signal driven by the circadian clock, but it only triggers ovulation every five to six days when the eggs in their ovaries are mature enough. I discovered that

I could switch on the estrous cycles of young females that hadn't yet started ovulating by giving them 16 hours of light each day. This sensitivity to day length helps explain why the population on the island started breeding about the same time in November each year (give or take a week or so).

The story of this particular population of kiore gives me pause for reflection on our attitudes to the ecosystems in which we live. Māori, and presumably kiore, had inhabited Tiritiri Matangi from at least the fourteenth century.[1] In the 1850s, European farmers arrived and introduced livestock grazing, which had recently ceased when I started my research in 1977. In 1984, a community habitat restoration project was launched that included extensive planting of native flora and the extermination of introduced species including kiore (New Zealand has no native land mammals except bats). The island is now a wonderful predator-free native wildlife sanctuary that is also actively protected from environmental damage caused by humans.

In contrast to the nocturnal kiore, we humans are programmed by our internal circadian timekeeping system to sleep at night. Our night vision is not great and when we are asleep, our brain largely disengages from what is going on in the environment. This makes us vulnerable so we seek shelter at night, which would also have been a good way for our distant ancestors to avoid nocturnal predators.

The circadian timekeeping system in mammals like us is relatively well understood and is covered in more detail in Chapter Two. Briefly, we have a circadian master clock consisting of about 20,000 nerve cells clustered together in the hypothalamus area of the brain. This master clock coordinates our circadian rhythms from the cellular level through to our moods and behaviour. It has two major tasks: to keep all our rhythms synchronised internally, and to keep them in step with the day/night cycle. Its most important time cue from the environment is the intensity of blue light, which it tracks via specialised cells in the retina of the eye (these are not part of the visual system that enables us to see).

Having a built-in circadian timekeeping system prepares us for the predictable environmental changes of the day/night cycle. For example, if we sleep according to our natural rhythms, in the few hours before we wake up in the morning the circadian timekeeping system starts preparing us for the demands of being awake. Our core body temperature begins to rise ahead of the increasing metabolic demands of wakefulness. The ability of our blood to clot increases, ahead of the increased likelihood of accidental cuts after we wake up and become active.

We are unique among species in our determination to try to override the pattern of living dictated by our circadian timekeeping system. Working night shifts, flying across time zones, or staying up late on the internet disrupts our exposure to the natural time cues provided by the 24-hour day/night cycle. The recent human habit of living in orbit around Earth creates a particularly bizarre environment for circadian clocks – for example, crew members on the International Space Station experience 16 sunrises in every 24 hours.[2] All these activities disrupt the intricate synchrony among circadian rhythms throughout the brain and body. A common analogy is that the result is like all the players in a symphony orchestra doing their own thing, or following different conductors, instead of keeping in time with each other. The result is not music, but cacophony. The adverse consequences of circadian desynchrony for our health, safety, and well-being are a central theme of this book.

Lunar and tidal rhythms: legacy of the moon's rotation and orbit around Earth

From observing 'lunatics' to planting crops at specific phases of the moon and predicting the tides, humans have known for millennia that the cycles of the moon influence us and other living organisms. There are multiple geophysical cycles associated with the moon, and to add to the complexity we talk about them in terms of the 24-hour solar day. This section looks at the evidence for internal clocks in living organisms that are adaptations to the environmental cycles generated by the moon.

The moon orbits Earth every 27.3 days (roughly four weeks). It also rotates on its axis once every 27.3 days, so to an observer on Earth its face doesn't seem to change (this is known as synchronous rotation). The combined effect of the moon orbiting Earth every 27.3 days and Earth spinning on its axis every 24 hours means that the moon goes past any given point on Earth every 24.8 hours – the lunar day.

The moon's gravitational pull on the oceans causes simultaneous tidal bulges on opposite sides of Earth, resulting in two tide cycles every 24.8 hours (with about 12.4 hours between high tides). The sun also exerts a gravitational pull on the oceans. When the gravitational pulls of the sun and the moon are aligned, at the new moon and full moon, there are spring tides, with more extreme highs and lows. When their gravitational pulls are at right angles, seven days after the spring tides, there are neap tides, with more moderate highs and lows.

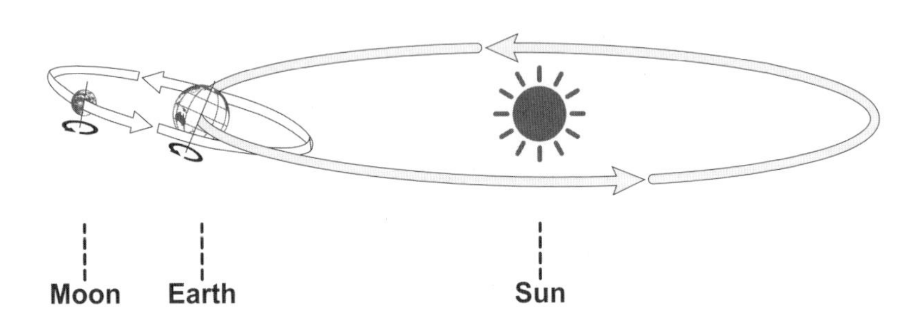

Figure 1.4 Orbits of the moon around Earth and Earth around the sun (not to scale). Both Earth and the moon also rotate on their axes.

The moon does not produce light – moonlight is reflected light from the sun. We see the lighted part of the moon getting larger (waxing) up to the full moon, and then smaller again (waning) down to the new moon. At new moon, the moon is directly between the sun and Earth, so the side facing us is away from the sun. The moon is directly overhead in the middle of the day and not visible at night. At full moon, Earth is between the moon and the sun. The moon is directly overhead in the middle of the night and reflects the sun's light back at us. The full cycle from one new moon to the next is about 29.5 days.

Light intensity is commonly measured in lux. One lux is equal to the illumination of a one-metre square surface that is one metre away from a single candle. Even at full moon, the intensity of moonlight as perceived by the human eye (under 1 lux) is much weaker than direct sunlight (up to 130,000 lux), office light (300–500 lux) or even your computer or phone screen (typically 30–50 lux). Nevertheless, behaviour patterns related to the cycles of the moon

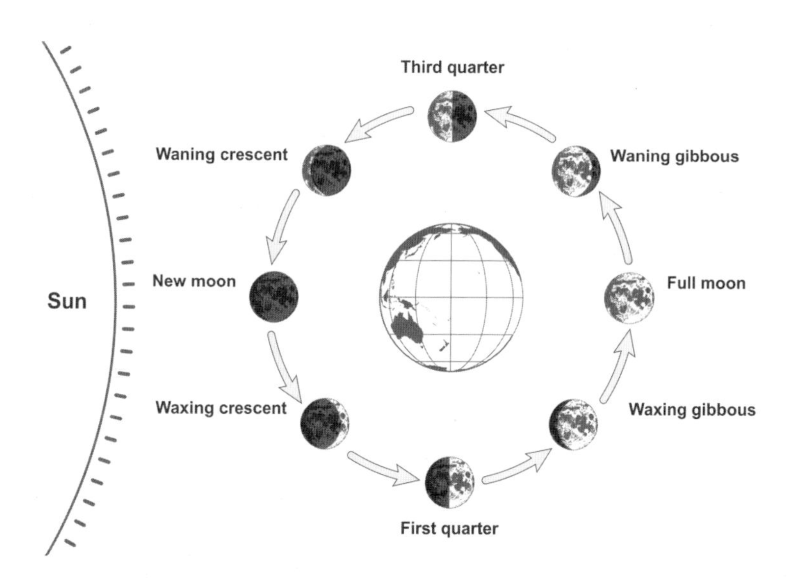

Figure 1.5 Phases of the moon as seen from Earth (not to scale).

have been documented in a wide range of organisms, particularly marine species. What causes them is often difficult to decipher. Some are apparently driven by internal clocks with cycle lengths close to the geophysical cycles (circatidal and circalunar clocks), while others seem to be a direct response to the environmental changes that accompany the tide cycles or the lunar month. In some species there is evidence to suggest that circadian and circalunar clocks interact.[3]

Ancient horseshoe crabs belonging to the genus *Limulus* provide an intriguing example of how complex rhythmicity can be in marine animals.[4] These unusual creatures have been around for about 450 million years. The circadian clock of horseshoe crabs greatly increases the sensitivity of their eyes to light at night, so they can probably see as well at full moon as they can during the day.

During the spring and summer months, females and males come ashore at high tide to breed. In the laboratory, where there are no tidal or lunar time cues, most horseshoe crabs continue to show increased physical activity roughly every 12.4 hours, mimicking their pattern of coming ashore at high tide on their home beach.

This suggests that horseshoe crabs have an internal circatidal timekeeping system that enables them to anticipate and synchronise their activity to the natural tidal cycles. When artificial tides are produced in the laboratory, the peaks in activity synchronise to them. This is the same principle as was shown earlier for the non-24-hour circadian clock of the kiore, which could be synchronised to an artificial 24-hour light/dark cycle.

In nature, the tide-related pattern of activity of horseshoe crabs is most obvious during the breeding season. It may help synchronise the behaviour of males and females to the local tides, thus enhancing the probability of finding a mate and of the female depositing eggs in an optimal location around the high-water line. Outside the breeding season, horseshoe crabs become much less physically active overall, but they have also been observed to move onto the tidal flats during high tides, when they are probably actively foraging, and then move off again during low tides.

Interestingly, in the laboratory horseshoe crabs are more active in warm water (17 degrees Celsius), regardless of whether the experimental light cycle is mimicking long summer days or short winter days. On the other hand, in cold water (4 degrees Celsius) the circatidal activity pattern (becoming more active about every 12.4 hours) disappears. It is possible that horseshoe crabs also have an internal annual timekeeping system (a circannual clock) that regulates the timing of the breeding system, but more research is needed to confirm this.

Do humans have internal clocks to predict and prepare for the cycles of the moon? Opinions vary. The 28-day menstrual cycle approximates the 27.3-day orbit of the moon around Earth so closely that we often assume that the two are related. However, this has not yet been scientifically confirmed. It could be a chance coincidence (badgers also have a 28-day menstrual cycle). In humans, the timing of births has also been thought to relate to lunar cycles. A study in New York City in the 1940s and 1950s found that birth rates were 2–3 per cent higher than average around the full moon, and 2–3 per cent lower than average around the new moon. More recent studies have not always replicated these findings. This could be due to changes we have made since the 1950s, such as increasing use of techniques to induce labour, elective caesarean birth, private clinics having fewer births on weekends, and changes in our use of artificial light.[5]

Intriguing evidence that men may have internally generated circalunar (monthly) rhythms was found in the Mars105 and Mars520 space flight simulation studies.[6] In these studies, 12 healthy young male volunteers spent 105 and 520 days, respectively, in an enclosed habitat designed to mimic living conditions during a round trip to Mars. This consisted of hermetically sealed interconnecting modules with constant environmental conditions, and the crews lived and worked like crews on the International Space Station (but not in microgravity). Although there were no monthly time cues, during the missions they had monthly rhythms in the total amount of sodium in their bodies, even though they had a constant daily intake of sodium and their body weight and total extracellular water content remained unchanged. Monthly rhythms were also detected in the levels in their urine of two hormones made by the adrenal glands, aldosterone and cortisol, which could have contributed to the monthly changes in total-body sodium.

There is a long-standing belief in a connection between lunar cycles and mental illness. According to Shakespeare,

> It is the very error of the moon. She comes more nearer Earth than she was wont. And makes men mad. (*Othello*, 5.2.135)

Indeed, a link between the moon and mental health is implicit in the word 'lunatic'.

How did this belief arise?[7] Some argue that before artificial light became readily available, people were more likely to go out and about on nights around the time of the full moon to work, hunt, travel, etc. This periodic sleep deprivation could switch people with bipolar disorder into mania or increase the rate of epileptic seizures. Lunar cycles could potentially also affect our mental health if they have a direct effect on sleep duration or quality, thereby changing waking mental function. There is some evidence to support this, with indications that there may be differences in how much the lunar cycles affect the sleep of men versus women and across the lifespan.

There is still a widespread belief that mental health crises are more common around the full moon, although recent studies of patterns of patient admissions do not support this.[8] There may be a methodological problem here. If we do have internal clocks changing how we function across the phases of the moon, they would need consistent exposure to the phases of the moon to stay synchronised to them. Society still works predominantly on the 24-hour solar day and most of us have variable exposures to artificial light at night, not regular exposure to moonlight. If some people have internal monthly clocks that affect their mental health, those internal clocks are unlikely to be running in step with the phases of the moon or synchronised among individuals. This makes it highly unlikely that studies focusing on populations of people would find consistent monthly patterns in sleep or any aspect of waking mental health, particularly among urban dwellers in industrialised countries.

In contrast, when US psychiatrist and sleep researcher Tom Wehr looked back at the switches in mood of 18 individual patients with bipolar disorder, he observed that the switches tend to be clustered in patterns lasting one, two, or three semilunar cycles (semilunar – about 14 days, the pattern of the spring and neap tides). This might suggest a possible influence of the moon's gravitational field.[9] Analysis of one patient's sleep/wake cycles suggested that both the 24-hour solar day and the 24.8-hour lunar day were influencing his sleep/wake cycle. When the 24.8-hour lunar cycle dominated, he switched from depression into mania with the new moon, and from mania into depression with the full moon. When he adhered to a rigid 24-hour schedule of rest and sleep during long periods of darkness every night, the lunar component in his sleep/wake cycle disappeared and his mood cycling stopped.

Clearly, there is a lot more to learn about how the cycles associated with the moon affect us and the rest of life on our planet.

Seasonal rhythms: legacy of Earth's rotation around the sun

In many cultures through the ages, the seasons have been celebrated as the repeating cycle of life: from birth in spring, through flourishing in summer, aging in autumn, and death in winter, followed by rebirth in spring.

Figure 1.6 The Green Man, a widespread traditional symbol of the regeneration of nature in spring. Artwork and photo: Philippa Gander

In chronobiology the perennial question arises: are the seasonal changes seen in many living organisms simply a direct response to the changing environmental conditions as Earth orbits around the sun, or is there an internal timekeeping system in those organisms that drives their seasonal changes and is sensitive to the environ-

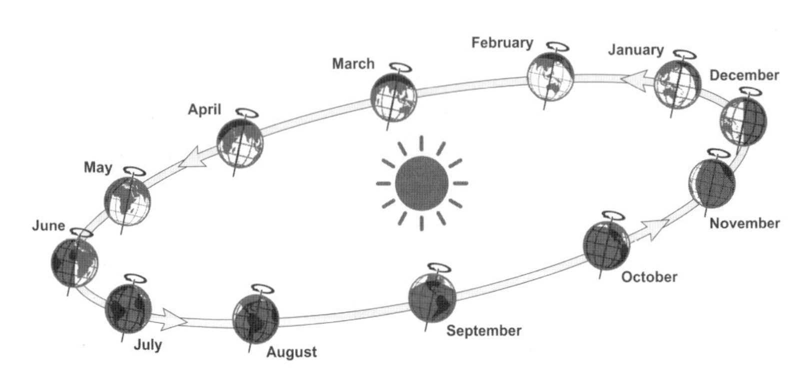

Figure 1.7 Annual orbit of Earth around the sun (not to scale).

Table 1.1 Seasonal Patterns Resulting from Earth's Tilt on its Axis

	Southern hemisphere	Northern hemisphere
December–February Southern hemisphere tilted towards the sun	Summer solstice and longest days in December	Winter solstice and shortest days in December
March–May Sun shines equally on both hemispheres	Autumn equinox in March	Spring equinox in March
June–August Northern hemisphere tilted towards the sun	Winter solstice and shortest days in June	Summer solstice and longest days in June
September–November Sun shines equally on both hemispheres	Spring equinox in September	Autumn equinox in September

mental changes? This section looks at the evidence for internal circannual clocks.

Earth takes 365.25 days to complete one orbit around the sun. To keep the calendar in step with this, we have three years of 365 days followed by a leap year with 366 days (by adding February 29).

Earth has seasons because its axis is tilted with respect to the plane of its orbit around the sun, which also means that the seasons in the northern and southern hemispheres are always opposite to each other, as summarised in the table above.

The full story is actually a bit more complex than this table suggests because changes in daylight hours across the year vary by latitude. At the equator (latitude 0 degrees), daylight lasts about 12 hours all year round. At the poles (latitudes 90 degrees north and south) there are almost six months of total darkness and six months of continuous sunlight per year. People living above 66.5 degrees (north or south) experience 24 hours of continuous daylight around the summer solstice and 24 hours of continuous darkness around the winter solstice. In Wellington, New Zealand, where I live (latitude 41.3 degrees south), seasonal daylength varies from just over nine hours to just over 15 hours. In temperate latitudes like Wellington,

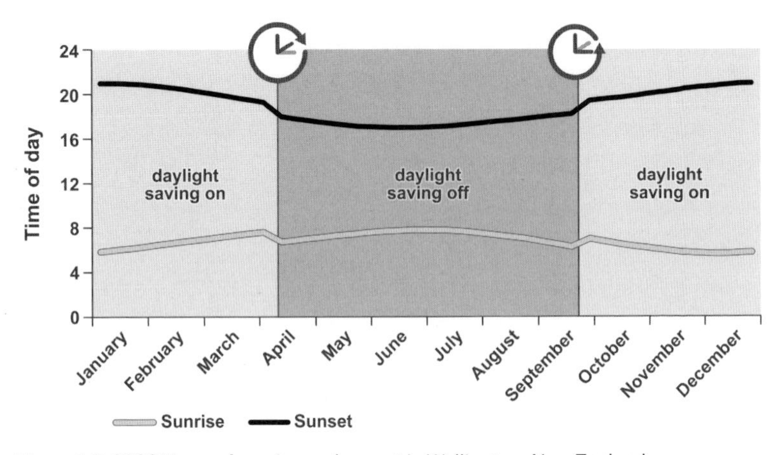

Figure 1.8 2020 times of sunrise and sunset in Wellington, New Zealand.
Data from the Royal Astronomical Society of New Zealand, https://rasnz.org.nz/in-the-sky/sun-rise-and-set

changes in daylength (photoperiod) provide a highly predictive signal of seasonal changes in the environment.

Do living organisms have internal circannual clocks that enable them to predict and prepare for the seasons? To demonstrate that this is so, you have to be very patient (and have long-term research funding). You need to keep an organism for years in an environment without seasonal time cues – constant lighting, temperature, and humidity. Nevertheless, a wide variety of species have been shown to have circannual clocks.

For example, in most breeds of sheep, ewes living in natural seasonal cycles come into breeding condition as the days begin to shorten in late summer–autumn.[10] Under these conditions, all the ewes in a flock begin ovulating at around the same time and the lambs are born the following spring. However, in an experiment where ewes were kept for four years in artificial short days (eight hours of light per 24 hours), they continued to show a rhythm of coming back into breeding condition, but this occurred on average every nine months, with some variability among the individual animals. Together with a variety of other evidence, this suggests that ewes have an internal circannual clock with an innate rhythm of about nine months that is synchronised by the 12-month cycle of the seasons. This is once again the same principle shown earlier for the non-24-hour circadian clock of the kiore, which could be synchronised by an artificial 24-hour light/dark cycle.

The idea that a single cell could contain a molecular mechanism that produces a cycle as long as a year initially seemed far-fetched. But fact can be stranger than fiction. The marine algae *Alexandrium* has been doing this for about 200 million years.[11] In summer, it grows as individual cells that self-replicate (by splitting into two identical cells) and can produce the massive toxic algal blooms known as red tides, which cause paralytic shellfish poisoning.

In autumn, pairs of cells join together to form a dense, resistant cyst that sinks to the cold darkness of the deep ocean floor for the winter, where there are no seasonal time cues. As spring approaches on the surface, the cysts germinate and split into two single cells that go back up and start the cycle again. Cysts collected in sea-floor sediment can survive in the laboratory for years in cold, dark conditions. When they are warmed up and placed in constant light, they show an 11-month cycle of reactivation, without any annual time cues from the environment.

Given that ancient single-celled organisms like *Alexandrium* have internal circannual clocks, it has been suggested that this might be an ancestral trait that was passed on to other organisms that arose later in evolutionary history.[12] Consistent with this idea is the evidence that a wide diversity of organisms have internal circannual clocks, including mosses, flowering plants, insects, molluscs, fish, reptiles, birds, and mammals, as well as algae.

One adaptive advantage of a circannual clock is that it provides an internalised sense of annual time where seasonal environmental cues are absent or ambiguous – for example, during hibernation when mammals hide away from the environment to escape the extreme conditions of winter, or when birds migrate seasonally, covering vast distances to change hemispheres and go from winter to summer. Another advantage is that many of the seasonal changes that different species must undergo to survive and reproduce require major physiological changes. It can take weeks to activate the reproductive system, grow or moult a winter coat, or gain weight ahead of migration or hibernation.

Compared to circadian timekeeping, we don't yet understand as much about how circannual timekeeping works in different organisms. For mammals, it has been proposed that there is a hierarchical circannual timekeeping system rather like the circadian timekeeping system for rhythms synchronised by the day/night cycle.[13] The circannual master clock probably resides in pacemaker cells in the pituitary gland and in the adjacent lining of the third ventricle (a fluid-filled space in the middle of the brain).

At the cellular level it is possible that the circannual clock genes may get switched on and off, which leads to marked differences in physiology and behaviour between summer and winter. For example, when mammals hibernate, they go through long periods of torpor, when their core body temperature can go as low as 0 degrees Celsius and their metabolism drops to between 1 and 5 per cent of normal. These are very radical changes compared to how they function when they are not hibernating. A possible mechanism for this has been found in adult thirteen-lined ground squirrels.[14] Not all genes are active all the time in all cells of the body. It has been shown that when these ground squirrels are in torpor, there is a shutdown in certain genes in the liver and skeletal muscles and in the production of the proteins that they code for. At the same time, other genes are activated, along with the production of their corresponding proteins. All this has to be switched back again for the ground squirrels to emerge from hibernation in spring. Clearly more research is needed on how the brain generates circannual rhythms – this research is fascinating, but it cannot be fast.

The primary environmental time cue that synchronises circannual clocks is the seasonal change in daylength.[15] Sensitivity to changing daylengths, known as photoperiodism, is found in many animals, plants and other types of organisms, although the underlying mechanisms vary.

In mammals, for instance, the critical translator of changing daylength to the body is the hormone melatonin. To understand how this might work we need to look more closely at some of what melatonin does in the body. Melatonin is one of two main 'messenger' hormones (the other is cortisol) that enable the circadian master clock to coordinate rhythms throughout the body. In the late evening, the circadian master clock switches on synthesis of melatonin in the pineal gland and then switches it off again in the early morning. This pattern of melatonin secretion is the same in species like humans who sleep at night and species like rats that are active at night, so melatonin is not always 'the sleep hormone', as is sometimes claimed. The circadian master clock not only switches melatonin synthesis on and off, but also has melatonin receptors, so it receives feedback about circulating melatonin levels.

The circadian master clock is not the only thing that switches melatonin synthesis on and off in mammals. Laboratory studies have shown that light above about 350 lux also stops us making melatonin. A recent Australian study with 62 healthy adults found that nearly half of their homes had light in the three hours before bedtime that was bright enough to suppress melatonin synthesis.[16] The study also confirmed that there are marked differences between individuals in the amount of suppression caused by light. Nevertheless, in this study the light levels in the average home would suppress melatonin by nearly 50 per cent in the average person. Greater exposure to light in the evening was associated with increased wakefulness after bedtime.

There are complex feedback loops in action here:

- The circadian master clock controls activity patterns and so influences the patterns of exposure to natural light, which in turn can reset the rhythm of the circadian master clock.
- The circadian master clock regulates the synthesis of

melatonin, and the amount of circulating melatonin can in turn reset the rhythm of the circadian master clock.
- Bright light also directly blocks the synthesis of melatonin.

As a result of all this, under natural lighting conditions at temperate latitudes, more melatonin is secreted in the brains of mammals during long winter nights than during short summer nights. This can have different effects, depending on the species. For example, some mammals breed during short days while others breed during long days, depending on the optimum times for pregnancy, giving birth, and the survival of offspring.

Based on current evidence, synchronisation of the mammalian circannual master clock by the seasonal changes in daylength appears to be a two-step process. First, the changes in daylength alter the amount of circulating melatonin via two mechanisms: 1) daylight directly suppresses melatonin synthesis; and 2) daylight synchronises the circadian master clock, which switches melatonin synthesis on and off. Second, seasonal changes in circulating melatonin levels synchronise the circannual master clock. The proposed circannual master clock cells in the pituitary gland are sensitive to melatonin and they control the release of a thyroid-stimulating hormone, which is critical for seasonal rhythms in many species.

So what happens to the circannual rhythms of mammals living at high latitudes, where the sun stays below the horizon continuously for many weeks in midwinter and continuously above the horizon for many weeks in midsummer? Do they have to rely on the changes in daylength around the spring and autumn equinoxes to keep their internal circannual clocks synchronised to a yearly cycle?

Reindeer living in Tromsø, Norway at 70 degrees north don't hibernate to avoid winter. Interestingly, they also don't have circadian rhythms in melatonin levels for about two months around the winter solstice and again for about two months around the summer solstice.[17]

One set of experiments looked at what happened to their circannual rhythms when they were transferred from natural day/night cycles into indoor environments where the temperature was held constant. From midwinter onwards, one group lived in continuous darkness and a second group lived in continuous light. Both groups showed an accelerated onset of the typical springtime increase in food intake, antler growth, and moulting of their winter coats.

The researchers concluded that this happened because in both protocols, the reindeer missed out on the increasing daylengths that they would have experienced outdoors in springtime. They interpret this as evidence that the internal circannual clock in these reindeer had a natural rhythm shorter than a year and that the increasing daylengths in spring somehow stretched it out to 12 months. One possibility is that when the circadian melatonin rhythm first returns, the days are still short and the nights are long, so relatively large amounts of melatonin are produced, which acts as a brake on the usual springtime changes driven by the circannual clock. As the days get longer, the nights get shorter, and the amount of melatonin produced decreases to a level where the springtime changes can begin.

Seasonal changes in daylength are not the only important time cue for circannual rhythms. For example, a great deal of energy can be required for migration, staying warm in cold winter temperatures, and breeding and rearing of the young. For some species, seasonal food availability can affect the timing of circannual rhythms.[18] Interestingly, the proposed circannual master clock cells lining the third ventricle of the brain are sensitive to the nutritional status of mammals.

We have the same basic biology as other mammals. Does this mean that we have retained similar adaptations to Earth's orbit around the sun, including a circannual clock that is sensitive to seasonal changes in day length? One key difference is that we have an additional strategy for managing seasonal changes – we modify our environment.

Historical and experimental evidence suggests that our modification of the physical environment since the industrial revolution has suppressed our seasonal responses. Our increasing use of artificial light has clearly changed our exposure to seasonal changes in photoperiod. Minimising seasonal changes in temperature (through heating and air conditioning) and food availability (through global food distribution and supermarkets) may also have played a role. Facilitated by the internet, our recent headlong rush into 24/7 living is also likely to be having an impact not only on our exposure to the seasonal changes in photoperiod, but also on the day/night changes in light intensity that are the main environmental time cue for the internal circadian clock.

Given strong evidence that breeding cycles in many mammals are controlled by an internal circannual clock, researchers have looked for evidence of this in human populations. Changes in conception cycles around the world are often cited as an example of how we may have altered our seasonal responses. There used to be spring and winter peaks in conception rates in temperate latitudes. With increasing industrialisation in different countries, the amplitude of the annual cycle in conception rates has declined and the autumn/winter peak has become increasingly dominant.[19] For example, in the USA the spring peak was larger until the 1930s, when it was overtaken by the winter peak. Looking at how these two peaks have changed over time, German chronobiologist Till Roenneberg has proposed that industrialisation results in many more people working indoors, thus reducing their exposure to seasonal changes in daylength and temperature. As with the breeding cycles of ewes kept for four years in short daylengths, if many people have circannual clocks that are no longer synchronised by the seasonal time cues and each individual is running on their own internal circannual rhythms, then they don't stay synchronised with each other and the seasonal pattern across the population diminishes.

There is also evidence, particularly in remaining pre-industrial populations, that the seasonal peaks in conception can be influenced by temperature, food availability, and/or cultural practices.[20] For example, the seasonal activities of Copper Inuit people living in Holman, 480 kilometres north of the Artic Circle, were documented in field studies in 1978–80.[21] During winter, people lived together in the settlement. Trapping and hunting activities, which involved men being out of the settlement for several days at a time, declined. Socialising increased with more visiting and shared festivities for Christmas and New Year. In contrast, during spring and summer, nuclear families and young couples spent greater periods of time outside the settlement in ice-fishing or seal-hunting camps, where they also had more privacy and opportunity for intimacy. Most conceptions occurred in spring and summer, and most births in the first half of the year.

Many other aspects of human physiology, behaviour and health also show seasonal patterns. For example, the brain functioning of healthy young people has recently been shown to vary seasonally.[22] The study had 28 participants and used functional magnetic resonance imaging (fMRI) to monitor the activity in different parts of their brains while they did a variety of tasks at different times of year. The patterns of seasonal variation were different for different brain functions. In regard to mental health, in both the northern and southern hemispheres there are winter peaks in attention deficit hyperactivity disorder, anxiety, mania in bipolar disorder, depression, eating disorders, obsessive compulsive disorder, schizophrenia, and suicide.[23]

Seasonal patterns in gene expression relating to the production of more than 4000 proteins have been documented in white blood cells and adipose (fat) tissue, and these patterns are reversed in people living in the northern versus the southern hemisphere.[24] The cellular composition of blood also varies seasonally, with reversed patterns in the northern and southern hemispheres. In the immune system, risk biomarkers for cardiovascular, psychiatric, and autoimmune diseases have been shown to peak in Europe in winter.

Once again, the fundamental chronobiological question is what causes all these seasonal patterns in our health and functioning. Multiple factors could be contributing to each pattern. Evidence for the importance of photoperiod is increasingly compelling. At the population level, the large British Biobank study of middle-aged adults controlled for many potential confounding factors and still found a significant association between daylength and cardiovascular mortality and risk factors (blood pressure, markers of inflammation, and body mass index), which all peaked in winter.[25] Longer winter nights were also associated with later and longer sleep, as well as higher rates of reporting insomnia and feelings of low mood and anhedonia (having little interest or pleasure in doing things). For many people with seasonal affective disorder, supplementing their winter light exposure using bright artificial light is an effective way to reduce their depressive symptoms.[26]

Does nightly melatonin production still vary seasonally in humans? Small experimental studies have found that people living in natural light/dark cycles (while camping at about 40 degrees north) produced melatonin for longer on long winter nights than on short summer nights.[27] This suggests that seasonal changes in melatonin might provide a synchronising cue to an internal circannual clock, if we have one.

Clearly, we still have a great deal to learn about circannual rhythms.

This chapter provides just a tiny glimpse into the intricacy and diversity of biological adaptations to the geophysical cycles on our planet that are found in all cell-based organisms. For me, 40-plus years of research in chronobiology has highlighted two vital themes. First, the more we learn, the more there is to know. Second, knowing more about the complex temporal adaptations of life on Earth has become increasingly urgent for our future as a species and for the future of the complex ecosystems that sustain us.

We have also forgotten much that our ancestors understood from careful direct observation of the cyclical changes in their physical and biological environments. This knowledge was often central to their spiritual beliefs and essential for their survival. For example, the ruins of the ancient Mayan city of Chichén Itzá in the northern Yucatán jungle are more than a thousand years old. In the middle of the site stands the 30-metre-high Pyramid of Kukulcán, the Feathered Serpent God. Around the spring and autumn equinoxes, the setting sun falls on the edges of the pyramid's stepped terraces, casting a series of interlocking, triangular shadows that create a beam of light slithering down the pyramid's sacred staircase like a giant snake. As the sun sets, the light beam connects with a stone serpent head at the base of the sacred staircase.[28] The Mayans clearly had detailed knowledge of seasonal changes in photoperiod as well as remarkable architectural and engineering skills.

Figure 1.9 The Pyramid of Kukulcán, Chichén Itzá, at the spring equinox.
Photo: Shawn Christie

Chapter Two

How Our Bodies Keep Time across the Day/Night Cycle

Circadian rhythms are a genetically programmed legacy of living on our rotating planet. We are generally unaware of them, except for our daily patterns of sleeping and waking. Nevertheless, they are affecting how we function and feel every minute of every day of our lives. Since the onset of the industrial revolution around the middle of the eighteenth century, more and more of us have adopted lifestyles where we do not, or cannot, live regular 24-hour days with enough sleep at night to stay fully functional and healthy.

This chapter introduces what we currently know about how the diverse circadian rhythms throughout our bodies are coordinated internally, and how our circadian clocks are influenced by 24-hour environmental time cues to keep them in step with the day/night cycle. Understanding how this circadian timekeeping system works provides the background for the subsequent chapters exploring the unprecedented challenges created by 24/7 living.

Early steps

Research in human chronobiology only really got under way around the middle of the twentieth century. Initially the focus was on whether we have an internal circadian clock, or clocks. Several intrepid early chronobiologists conducted experiments on themselves by living in caves for extended periods. The idea of these time isolation experiments was to strip away any 24-hour environmental cycles that might be driving our daily rhythms.

In 1962, Michel Siffre, a 23-year-old French speleologist and chronobiologist, spent two months alone in the Scarasson cave 130 metres under the French Alps. He lived in a world without 24-hour time cues – constant temperature and humidity, artificial light that he controlled, and one-way communication with the ground crew on the surface whom he called when he woke up, when he ate, and just before he went to sleep. Like the kiore in time isolation in Chapter One, Siffre's internal sleep/wake cycle was longer than 24 hours, averaging about 24 hours and 30 minutes. The fact that he did not spontaneously live on a 24-hour schedule was seen as evidence that his daily activity patterns were not being driven by (unknown) subtle 24-hour time cues from the environment. It showed that indeed humans do have endogenous circadian clocks. When Siffre emerged from the cave, he thought he had lived fewer days than had actually occurred above ground.[1]

Since the 1960s, chronobiologists have built time isolation laboratories that are safer and more comfortable than caves. In these laboratories all 24-hour time cues are excluded, and the researchers can impose different routines. For example, like Siffre alone in his cave, research participants can be given complete control of the artificial time cues in the laboratory. They can choose when to turn the lights on and off, when to go to bed and get up, when and what to eat, and when to do any other waking activities available to them.

Alternatively, participants can be scheduled to live a 28-hour

Figure 2.1 Michel Siffre emerging from two months alone underground after his first experiment in time isolation in Scarasson cave in 1962.
Photo: C. Sauvageot, https://journals.openedition.org/dynenviron/567

'day', divided like the 24-hour day into one-third for sleep and two-thirds for waking activities. When the researcher turns the lights off, participants must be in bed in the dark trying to sleep for nine hours and 20 minutes. When the researcher turns the lights on, participants must be awake and active for 16 hours and 40 minutes, during which regular activities are scheduled, including three meals and various performance tests. The circadian master clock cannot synchronise to a 28-hour 'day', so participants on this schedule are forced to try to sleep at different times across their circadian master clock cycle.

By using a range of different protocols, researchers have been able to tease apart the characteristics and components of the internal circadian timekeeping system. Jürgen Aschoff and Rütger Wever were notable pioneers with a laboratory constructed in an underground bunker in the beautiful village of Andechs near Munich, which was a place of pilgrimage when I was setting out as a young chronobiologist.[2] Over 400 studies were conducted in this laboratory between 1964 and 1989.

Time isolation laboratories continue to be developed and provide rich new information and ideas about how the internal circadian timekeeping system works. However, daily life is much more complex than the set routines and lack of competing demands that characterise life in a time isolation facility. Monitoring the rhythms of people going about their daily lives required the development of new technologies.

Monitoring rhythms in daily life

Technological advances were opening a whole new world of research possibilities when I moved to the Silicon Valley in 1983 to join the NASA Fatigue and Jet Lag Program at Ames Research Center in Mountain View. The programme had Congressional support because of concerns raised in 1980 about whether new knowledge about the human circadian clock and sleep had implications for aviation safety. When I arrived, they were conducting large field studies with airline pilots, using the first generation of 'portable' monitoring systems that could continuously measure the circadian rhythms of people going about their daily lives. I also got my first personal computer.

Joining the NASA team proved to be a crucial turning point for me as a scientist. I had to move from basic science experiments and biomathematical modelling in the tidy, controlled laboratory world to researching and applying science in the complex and largely uncontrollable real world. I worked with a team of extraordinary innovators, colleagues, and mentors, notably Curt Graeber, John

Lauber, and Charles (Charlie) Billings. I vividly remember Charlie telling me that my work now had a constituency – the US taxpayers who paid my salary. My work had to be relevant to them, and I also had to be able to explain it to them. This guidance provided the foundation for the rest of my career.

For three initial studies, the NASA team recruited 140 pilot volunteers flying scheduled commercial long-haul, short-haul, and night-cargo trips. Their activity patterns, heart rate, and core body temperature were continuously monitored before, during, and after trips. This meant continuously (almost) wearing an activity monitor on their non-dominant wrist, three chest electrodes, and a rectal probe. Leads from the sensors were threaded through their clothing and attached to a recording device, the 'Vitalog', which was worn in a belt pouch. It was state-of-the-art at the time. To demonstrate how unobtrusive it was, I once wore it for 35 days, including on a mule ride up and down the walls of Yosemite Valley.

In addition to wearing a Vitalog, the pilots were asked to rate how they were feeling every two hours while they were awake. Fatigue was rated from 0 (most alert) to 100 (most drowsy). Mood was rated on 26 adjectives from 0 (not at all) to 4 (extremely), from which three aspects of mood were analysed: positive affect, negative affect, and level of activation. Apart from contributing to science and aviation safety, an incentive to encourage pilots to participate was a free pass to a NASA Space Shuttle launch.

The monitoring technology was very primitive and cumbersome by current standards, but it literally opened up a whole new era of understanding about the complexities facing the human circadian timekeeping system in the real world. Figure 2.2 illustrates the rhythms of a short-haul pilot before, during, and after three duty days where he was flying multiple short flights up and down the US East Coast and sleeping at night.[3]

The top panel shows that his core body temperature was increasing before he woke up in the morning. His body was being prepared

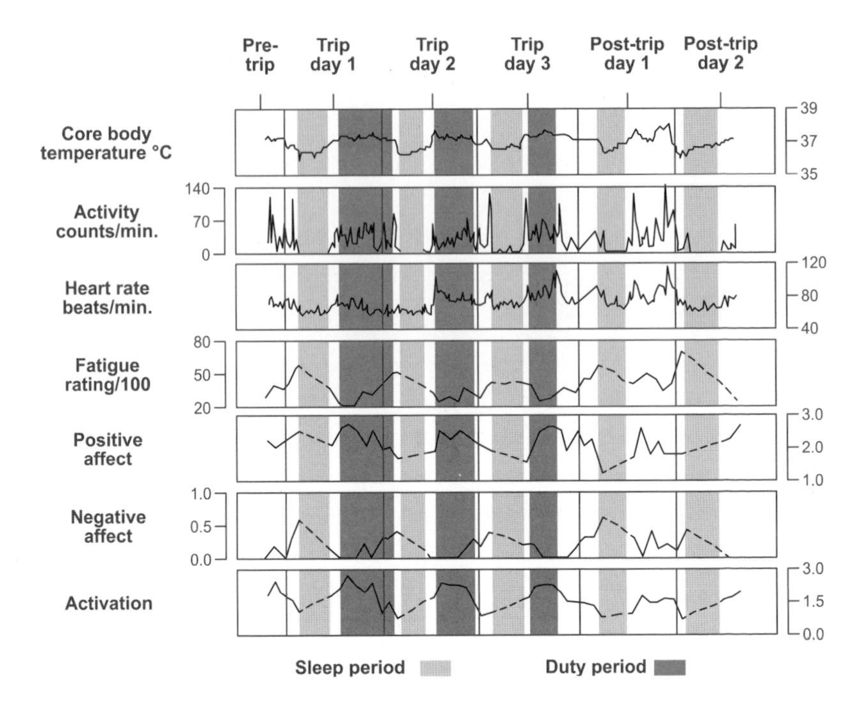

Figure 2.2 Rhythms of a short-haul pilot before, during, and after a three-day trip flying up and down the US East Coast. Vertical bars: dark shading – duty periods; light shading – sleep times recorded in a daily logbook.

for the additional metabolic demands of wakefulness. As expected, his heart rate was higher when he was more physically active and lower when he was less physically active (the second and third panels from the top in the figure). Nevertheless, there was still a detectable circadian rhythm in his heart rate after accounting for the effects of his changing activity levels. The bottom four panels in the figure show that he felt at his best (lowest fatigue and negative affect, highest activation and positive affect) a few hours after he woke up.

Although he was not consciously aware of it, this pilot's physiology and mood were changing systematically across the 24-hour day. His alertness and performance capacity would also have been changing but were not measured in these initial studies.

This and many other studies clearly demonstrate that all aspects of our functioning are changing systematically across each day of our lives. Does this have any practical relevance? Figure 2.3 demonstrates a situation where accidents are (unfortunately) common enough for the observed patterns to be reliable. It shows the timing of fatal and serious-injury crashes on New Zealand roads in the years 2017–19, in which driver fatigue was identified as a causal factor.

The early morning peak in fatigue-related crashes coincides with the time in the daily rhythm of the circadian timekeeping system when real-world and laboratory research predicts we are likely to be functioning at our worst. This is around the time of the daily minimum in core body temperature, when we feel sleepiest and our ability to perform different tasks is close to its daily low point. (The precise timing of the daily low point depends on the type of task being tested.) The afternoon peak in fatigue-related

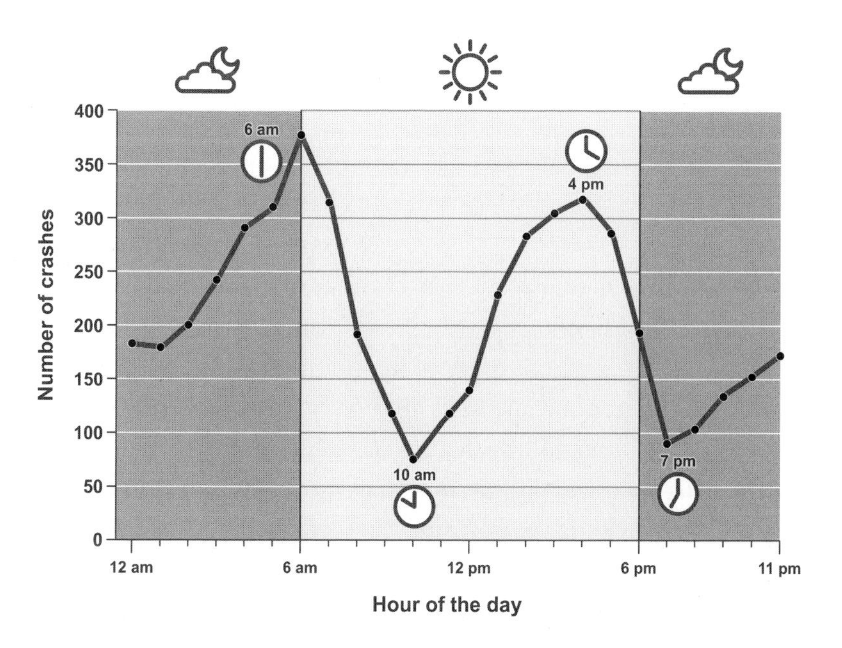

Figure 2.3 Timing of fatal and serious-injury motor vehicle crashes attributed to driver fatigue on New Zealand roads, 2017–19. Figure courtesy of Waka Kotahi New Zealand Transport Agency

crashes corresponds to the afternoon nap window, when sleepiness again increases, particularly if we have not had enough good-quality sleep in the last few nights. Regulation of sleep by the circadian timekeeping system is described in more detail in Chapter Four.

Finding the circadian master clock

By the 1970s, the hunt was on to locate the circadian master clock in mammals. Researchers were inserting electrodes into the brains of rodents to see the effects of selectively destroying different areas. In 1972, two bundles of neurons in the hypothalamus – together known as the suprachiasmatic nucleus (SCN) – were identified as the central pacemaker of the circadian timekeeping system, the master clock.[4] The SCN is so named because its bundles of neurons sit on top of the optic chiasm, where the optic nerves cross. Destroying them in animals isolated from environmental time cues resulted in complete loss of daily patterns in physical activity, feeding and drinking behaviour, hormone release, and body temperature.

We are diurnal primates, not nocturnal rodents. When I took up my first postdoctoral job at Harvard Medical School in 1980, the question was whether the SCN also functioned as the circadian master clock in humans. It had already been shown that lesioning the SCN in another diurnal primate, the South American squirrel monkey, resulted in loss of circadian rhythms in animals isolated from environmental time cues. However, there was still debate about whether humans even had an SCN. This was partly due to lack of research. Another confounding factor was that the bundles of neurons that make up the SCN sit on either side of the third ventricle, a fluid-filled space in the brain which is a different shape in humans versus rodents. In a pivotal 1980 paper, a group led by Ralph Lydic from the Harvard laboratory reported on a post-mortem study of human brains which identified small, loosely packed clusters of neurons that subsequent research confirmed are indeed our SCN and function as our circadian master clock.[5]

Individual neurons from the SCN can be kept alive in a petri dish for days. The 'firing rate' of a neuron is the number of electrical signals it sends out per second. Normally, the pattern of firing by each neuron communicates information to the other neurons connected to it. Even when SCN neurons are completely disconnected from each other in a petri dish, each neuron continues to generate its own circadian rhythm in firing rate. Interestingly, SCN neurons fire faster during the day than at night in all mammals, whether they are diurnal or nocturnal. Being diurnal, nocturnal, crepuscular, or switching between different patterns of activity across the seasons is controlled in another area of the brain (the dorsomedial hypothalamic nucleus or DMH), which receives input from the SCN and integrates it with information on feeding, temperature, social, and other cues.

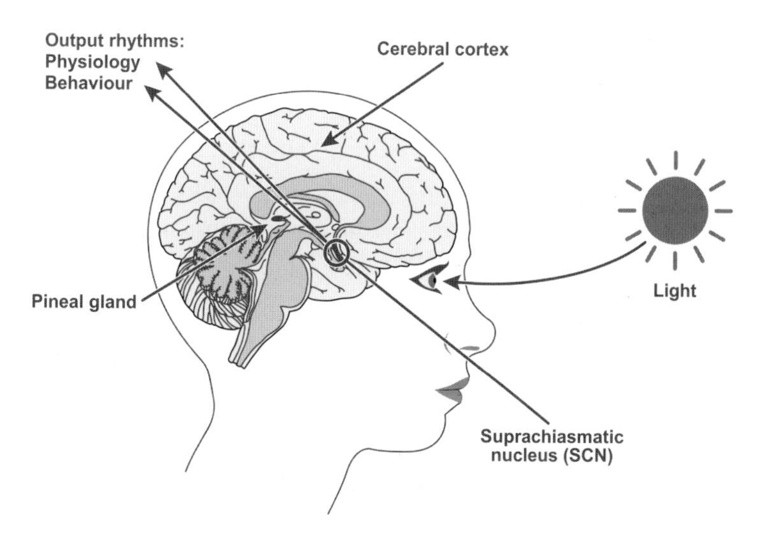

Figure 2.4 Location of the circadian master clock in the human brain.

Coordinating circadian rhythms in cells, tissues, and organs

Like all other cells in the human body, the SCN neurons contain the molecular circadian clockworks – the small number of clock genes that interact to generate circadian rhythms in the concentrations of clock proteins, with a cycle length of about 24 hours. In turn, the circadian rhythms in clock proteins drive circadian rhythms in a multitude of clock-controlled genes, resulting in cascades of circadian rhythms in the functioning of cells throughout the body. The combinations of genes that are clock-controlled vary among tissue types, enabling the unique functions of different tissues.

This vast network of cellular clocks has to communicate somehow, and act in synchrony to ensure that rhythms in different functions occur at optimum times in the day/night cycle.[6] The SCN neurons form a tightly coupled network that sends timing information to clocks elsewhere in the brain and in organs throughout the body. The SCN neurons can also stay in step with the day/night cycle because they receive light information directly from dedicated cells in the retina of the eyes. This is explored in more detail below.

When it was discovered that the SCN neurons received information about environmental light, it was thought that they were the top of a hierarchy of circadian control, and that they coordinated rhythms in everything else downstream. However, knowledge in this area is expanding fast. Rather than this 'dictatorship' model, current thinking is that the circadian timekeeping system is more like a 'federal' system, with interactions among different autonomous circadian clocks in the brain and the body. Figure 2.5 summarises this 'federal' idea. Each step is described as follows.

Step 1: The SCN master clock receives information about light intensity via a direct neural pathway from specialised cells in the retina of the eyes. These cells are not part of the visual system and they

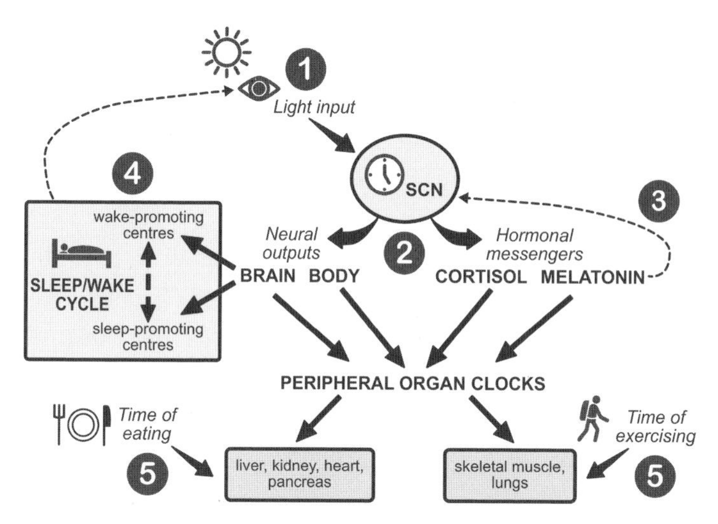

Figure 2.5 Diagram of the mammalian circadian timekeeping system.

contain a unique pigment, melanopsin, that makes them sensitive to blue light. Blue light, the colour of the sky on a clear day, has a role in synchronising the circadian timekeeping system in a wide variety of plants and animals.[7]

Step 2: The SCN master clock synchronises other circadian rhythms via two main types of output. It has a strong neural network that transmits time signals to other areas of the brain and body. It also regulates the production of two key 'messenger' hormones, melatonin and cortisol, which act on peripheral organs. In addition, the SCN modulates the sensitivity of the peripheral organs to these hormones.

Step 3: The SCN master clock itself has melatonin receptors, i.e., it is sensitive to changing levels of melatonin. This feedback is indicated by the dashed line on the right-hand side of the figure.

Step 4: Among its neuronal outputs, the SCN master clock has connections to brain areas that promote sleep and other brain areas that promote wakefulness. By modulating these competing brain areas, it influences the timing of sleep. When we sleep, we close our eyes (and generally choose a dark place to sleep), so our sleep/wake cycle modifies our exposure to environmental light. This feedback is indicated by the dashed line on the left-hand side of the figure.

Step 5: Some of the peripheral organs can function as autonomous circadian clocks (i.e., without input from the SCN). They normally receive light-based time signals from the SCN, but they can also receive other time cues independently of the SCN. For example, there are circadian clocks in the liver, heart, and pancreas that also get time cues from when we eat. Eating a hamburger in the middle of the night sends a message to these organs that conflicts with the message they get from the SCN saying that it is dark outside, and you should be sleeping and fasting. When peripheral clocks are reset by eating schedules, the SCN does not adapt to their schedule. Another group of peripheral organs, including skeletal muscle and the lungs, independently get time cues from when you exercise, which can conflict with the light-based time cues from the SCN.

Figure 2.5 highlights a fundamental reason why 24/7 living challenges our circadian timekeeping system. It can disrupt the intricate synchrony among different rhythms in the body. Rapidly expanding evidence supports the conclusion that this synchrony is essential for optimising our daily mental and physical functioning, safety, and well-being, and, in the longer term, our mental and physical health.

Individual differences in the circadian master clock

Harvard Medical School has a time isolation laboratory that has made a huge contribution to our understanding of the human circadian timekeeping system, led by another ground-breaking chronobiologist,

Charles Czeisler. To look at the innate rhythm length (period) of the SCN master clock in humans, the Harvard team compiled data from time isolation studies with 52 women and 105 men whose ages ranged from 18 to 74 years.[8] The studies tracked the circadian rhythms in core body temperature and melatonin as markers of the SCN's circadian rhythm in firing rate. When they were not synchronised to 24-hour time cues from the environment, the average length of their circadian rhythms was 24 hours nine minutes (shortest 23 hours 29 minutes, longest 24 hours 36 minutes). On average, the circadian rhythms of women were six minutes shorter than those of men, but there were no differences by age.

When they were living in a 24-hour day/night cycle, the participants with shorter circadian rhythms went to sleep and woke up at earlier times of day than those with longer circadian rhythms. Other studies have found mutations in the core circadian clock genes in some people with very short circadian rhythms and very early sleep times, as well as in some people with very long circadian rhythms and very delayed sleep times.[9]

Anecdotally, we all know people who prefer to sleep relatively early or relatively late. Different preferences for when we sleep are known as chronotypes. Across puberty teenagers typically become more evening-type, while across adulthood most of us gradually become more morning-type. Independent of these developmental and age-related changes, chronotypes reflect basic differences between people in their circadian timekeeping systems. Variations in the intrinsic length of the SCN master clock rhythm may explain some of these differences. However, the Harvard study did not find any differences between younger and older adults in the lengths of their circadian rhythms, so clearly there are other factors that influence when we prefer to sleep in everyday life.

There are various ways to measure chronotype. Two questionnaires have been widely used, with one focusing on preferred times for sleep and other activities, and the other on usual sleep timing on scheduled and free days. In large population surveys using these standardised questionnaires, most of us fall somewhere in the middle between extreme morning-types and extreme evening-types. Some survey studies, but not all, find that women tend to be more morning-type than men.[10] Smaller studies have monitored sleep timing and compared circadian rhythms between morning-types and evening-types. They confirm that scores on the questionnaires do relate to physiological differences in circadian timing. However, the scores also reflect the influence of other factors.

Synchronising the circadian master clock to the 24-hour day

A universal feature of circadian rhythms in everything from unicellular algae to humans is that internal rhythms are not exactly 24 hours, but they can be synchronised to 24 hours by time cues from the environment. In the Harvard studies cited above, the longest circadian master clock rhythm measured was 24 hours 36 minutes. This person would need to shorten their rhythm by 36 minutes every day to stay on a 24-hour pattern. The shortest circadian master clock rhythm measured was 23 hours 29 minutes. This person would need to lengthen their rhythm by 31 minutes every day to stay on a 24-hour pattern. How does this work?

When the dedicated cells in the retina are stimulated by light, they send a signal to the SCN that accelerates the firing rate of the SCN neurons. Depending on where the SCN is up to in its firing rate rhythm, this can have different effects, as described in Figure 2.6. The shaded vertical bars represent two hours of bright light exposure for a person who is living in constant dim light and deciding their own schedule in a time isolation laboratory. Because there are no 24-hour time cues, they are living on their internal circadian time dictated by the SCN master clock (known as free-running).

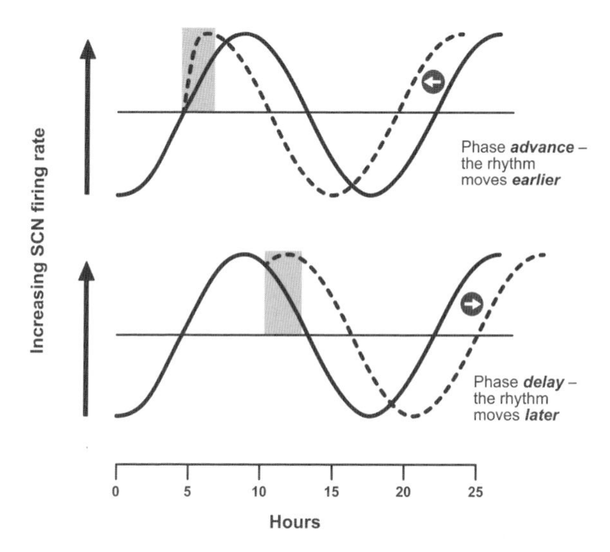

Figure 2.6 Effect of timed bright light exposure on a subject in a time isolation laboratory. Light can either advance or delay the SCN master clock rhythm, depending on when it is seen. The shaded vertical bar represents bright light exposure for two hours for someone living in constant dim light and controlling their own schedule (free-running).

The upper panel shows that if the light comes on while the SCN firing rate is increasing, it increases even faster. The peak firing rate occurs *earlier* than it would have without the light. Consequently, the whole rhythm moves earlier (to the left). This shortening of the SCN rhythm by a light pulse is known as a phase advance. The lower panel shows that if the light comes on after the SCN firing rate has started decreasing, the firing rate goes up again until the light stops. The firing rate then starts to decrease again, but *later* than it would have without the light. Consequently, the whole rhythm moves later (to the right). This lengthening of the SCN rhythm is known as a phase delay.

Translating this to everyday life, when you are synchronised to the 24-hour day/night cycle, light exposure in the morning causes phase advances (as in the upper panel) and light exposure in the evening causes phase delays (as in the lower panel). Light exposure in the middle of the day has minimal effect.

If you can recall staying up all night, you may remember the feeling of hitting a really low point in the early hours of the morning and then starting to feel better, even though you have been awake for longer. Around this low point, the effects of light on the SCN master clock switch from causing large phase delays to causing large phase advances. This is also the time when core body temperature reaches its lowest, when we feel sleepiest, and when we are reaching our daily low point in how well we can do many tasks. This has implications for productivity and safety during night shifts (discussed in Chapter Five).

The changing effects of light at different times in the SCN master clock rhythm can explain how Earth's 24-hour light/dark cycle is able to synchronise non-24-hour SCN master clocks. The person in the Harvard studies who had an SCN rhythm of 24 hours 36 minutes would need to shorten it (phase advance it) by 36 minutes every day to stay in synchrony with the 24-hour day/night cycle. Theoretically, they could do this by getting the right amount of bright light every morning. Conversely, the person who had an SCN rhythm of 23 hours 29 minutes would need to lengthen it (phase delay it) by 31 minutes every day. Theoretically, they could do this by getting the right amount of bright light every evening.

Of course, in everyday life synchronising to the 24-hour day/night cycle is not that simple. The size of the shift that light can cause at any given time in the SCN pacemaker cycle depends on how much light is hitting the receptor cells in the retina (more blue light results in a larger shift). However, there is a limit to this effect because if all the receptor cells in the retina are already activated, then adding more light has no extra effect. In addition, our daily lives are full of other time cues and constraints.

All the external time cues that can synchronise circadian clocks (the SCN master clock and the clocks in the peripheral organs) have the property of having different effects depending on when they occur in a clock's internal rhythm. For example, you may recall from

Chapter One that the SCN has melatonin receptors which make it sensitive to the level of circulating melatonin. Taking melatonin in the morning delays the SCN master clock's rhythm, and taking melatonin in the evening advances its rhythm, i.e., the opposite effects to light exposure.

Many blind people and some sighted people have difficulty synchronising to the 24-hour day. For most of these people, their sleep tends to drift later day after day (their SCN rhythm is longer than 24 hours). Sometimes their biological sleep time dictated by the SCN will coincide with night. At other times, their biological sleep time will be occurring during the day, and if they try to stay on a routine of sleeping at night, they are likely to experience insomnia, since they are trying to sleep when the SCN is pushing the brain into wake mode. In some cases, this can be corrected by increasing the strength of the 24-hour time cues. In principle this could be done by combining the effects of light exposure and melatonin. People who have an SCN rhythm longer than 24 hours could increase their light exposure in the morning and take melatonin in the evening. Those with an SCN rhythm shorter than 24 hours could increase their light exposure in the evening and take melatonin in the morning.

There has been rapid expansion of an industry producing synthetic melatonin and it is often promoted as 'the sleep hormone'. However, as we have seen already, melatonin is much more than a sleep aid. In mammals it: regulates seasonal breeding (see Chapter One); is a messenger hormone via which the SCN master clock regulates circadian rhythms in other parts of the brain and body; and can phase-shift the SCN rhythm in firing rate. The observation that melatonin levels are highest at night in both diurnal and nocturnal mammals may indicate that inducing sleep is not its primary function.

On the other hand, there is evidence that melatonin can improve sleep for some people, particularly in situations where they are trying to sleep out of step with their own circadian rhythm in melatonin, such as during shift work or jet lag. However, it is important to note that melatonin is not an effective treatment for all types of sleep disorders and there are important unresolved questions about its safety for certain groups, as well as about the effects of long-term use.[11]

Socially dictated time versus solar time

Having a light-sensitive circadian master clock gives us flexibility to adapt to changes in the natural day/night cycle, such as the gradual seasonal changes in photoperiod. We can also adapt to the abrupt changes in the natural day/night cycle that we have created with air travel across time zones.

Figure 2.7 shows my sleep patterns and light exposures when I flew from Wellington via San Francisco and Atlanta to Montréal in 2016, to give a plenary address at an International Civil Aviation Organization symposium on managing pilot fatigue. From 31 March to 1 April, my 24-hour day/night cycle advanced six hours and six months (from autumn in New Zealand to spring in North America). In the 30 years since the study with the pilots at the beginning of this chapter, there have been huge technological advances. I wore a small, watch-sized device (an actigraph) that recorded my wrist movements and average light exposures every minute. The activity data were run through a computer algorithm that decided for every minute whether I was awake or asleep. The reliability of this algorithm had been tested by comparing its minute-by-minute predictions against sleep measured using polysomnography – the gold-standard technology of monitoring brain activity, eye movements, and chin muscle tone during sleep – which is described in Chapter Four.

From my arrival in the new time zone, I tried to schedule my activities on local time. My sleep did not adapt immediately – the SCN master clock takes several days to become fully resynchronised to

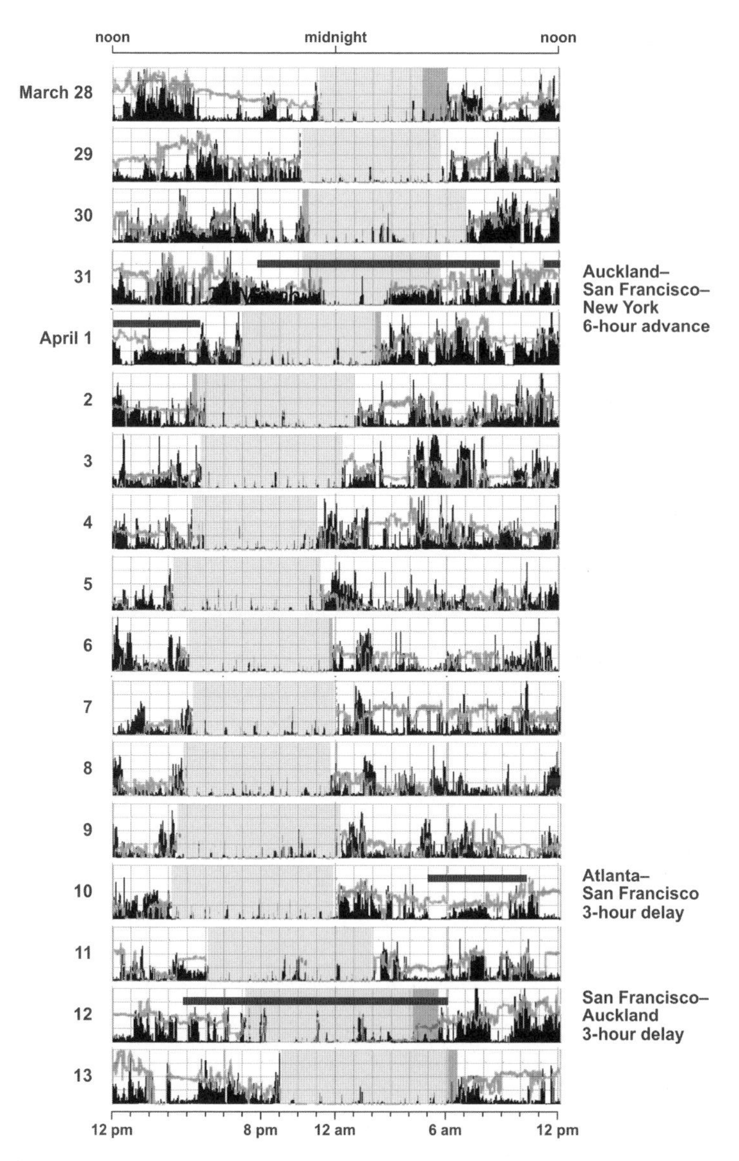

Figure 2.7 Sleep patterns and light exposures on a trip from Wellington, New Zealand, to Montréal, Canada (via San Francisco and Atlanta).

- Each horizontal panel represents one day, from noon to noon New Zealand standard time.
- The height of the black bars indicates the amount of wrist movement per minute.
- Lighter shaded boxes indicate when I was asleep; darker shaded boxes indicate when I was trying to sleep, but was awake.
- Dark horizontal bars indicate transmeridian flights.
- The continuous wiggly grey line indicates light levels.

a new time zone. But, having a light-sensitive circadian master clock, I was eventually able to shift over to Montréal time, and then back to Wellington time on my return home.

On the other hand, our light-sensitive circadian master clock becomes problematic when we try to override it with socially imposed time schedules that are at odds with the natural day/night cycle.

Time zones are one example of this. Meridians (or degrees of longitude) are imaginary lines that run through the North and South Poles and divide the planet into 360 segments (like the segments of an orange). In 1884, it was agreed that the zero meridian would run through Greenwich, near London, so the reference became Greenwich Mean Time (GMT), now Coordinated Universal Time (UTC). For every 15 degrees of longitude you travel east, solar time moves one hour later, so this was the theoretical basis for designating time zones.[12] However, time zones have morphed, usually for political reasons. I lived for some years in south-west France. Toulouse is only 1.5 meridians east of Greenwich, but it is on Central European Time, which is centred 15 meridians east of Greenwich (roughly through Prague). Social time in Toulouse is thus about 54 minutes ahead of solar time. Particularly in winter, the sun can seem to take an awfully long time to come up in the morning. China spans almost five solar hours but runs on solar time in Beijing, which is on its eastern edge. Consequently, if you are in the ancient Silk Road city of Kashi on the western edge of China, social time would be five hours ahead of solar time.

Daylight saving time (advancing the clocks one hour in the spring and delaying them one hour in the autumn) is effectively a decision to move social time in summer by one time zone to the east. However, solar time does not change. Introducing this one-hour discrepancy between social time and solar time has some surprisingly dramatic consequences. Short-term effects after the spring change to daylight saving include: shorter sleep; greater sleepiness in adolescents (which is consistent with sleep timing getting later across puberty);

increases in general accidents, visits to the emergency room, heart attacks and strokes; more negative mood; and a drop in the stock market.[13] There are contradictory findings regarding traffic accidents and hospital admissions, which may depend on 1) the time of year that daylight saving begins and ends, and 2) on latitude (since the day length changes with latitude).

What causes these effects? Limited evidence suggests that the SCN master clock does not always adapt fully to daylight saving time. For some people, sleep timing may adapt on scheduled days (typically workdays) but not on free days. People whose SCN master clock does not adapt fully to daylight saving time are at greater risk of losing internal synchrony between their different circadian rhythms, which could have detrimental effects on their health, safety, and well-being.

Social jet lag – the difference in sleep timing between workdays and work-free days – is another example of social time being out of step with solar time and the SCN master clock.[14] Two New Zealand studies highlight some of the causes and consequences of social jet lag. A 2008–9 national survey found that significantly more Māori adults (29.6 per cent) than non-Māori adults (23 per cent) reported at least two hours more sleep on work-free days.[15] This difference was partly explained by greater socioeconomic deprivation and more night work among Māori. Moderate/extreme evening chronotypes were also likely to report at least two hours of sleep extension on free days. When the 815 non-shift workers in the Dunedin longitudinal cohort study reached 38 years of age (in 2010–11), having a higher social jet lag score was associated with higher body mass index, greater fat mass, and increased likelihood of being obese and meeting the criteria for metabolic syndrome, after controlling for the influence of sex, chronotype, and sleep duration.[16] The authors concluded that further research on social jet lag could inform obesity prevention.

Shift work is an increasingly common social practice that sets up repeated bouts of conflict between social and solar cues for the SCN master clock, which rarely adapts fully to altered work schedules. For many people, shift work also creates tension between work demands and having the time and energy for essential life activities outside of work. Chapter Five focuses on the challenges of shift work, and some of the solutions that arise from applying chronobiology to reduce its harmful effects.

Although research in human chronobiology only really got under way around the middle of the twentieth century it is now clear that, like all cell-based life forms on Earth, we live with both the advantages and constraints of having an internal circadian time-keeping system. From the molecular level to the behavioural and emotional level, how we function changes systematically across the 24-hour day/night cycle.

Despite the dramatic increase in our knowledge in this area, we are continually devising new technologies and ways of living that conflict with our internal circadian timekeeping system. It seems to me that a major contributing factor is that we undervalue sleep – the essential third of life that most of us know least about (since we are not conscious at the time) but which is essential for the other two-thirds. Sleep is duly the topic of the next two chapters.

Although we have learned a great deal, I sometimes feel that the science of human chronobiology is permanently in catch-up mode. Nevertheless, I remain convinced that the knowledge we already have can make a constructive difference for many of us in our daily lives.

Chapter Three

Sleep – How We Know What We Know

One of my favourite paradoxes is that my waking brain has spent an awful lot of time trying to understand what my sleeping brain does without my conscious oversight. How do we 'know' what we 'know'?

This chapter illustrates a theme that recurs throughout the book. It uses our current understanding of sleep to illustrate how we prioritise waking knowledge gained through quantitative science, at the risk of overlooking other kinds of knowledge that are central to human experience and to our ability to resolve some of the big issues we are facing.

Two kinds of sleep

The beginning of the modern era of sleep science and medicine is often linked to the discovery of two different kinds of sleep, attributed to Eugene Aserinsky and Nathaniel Kleitman at the University of Chicago Medical School. In their paper in the journal *Science* in 1953, they described their experiments monitoring people sleeping right through the night.[1] Their method for monitoring sleep – polysomnography – has evolved since then but is still the gold standard. They monitored electrical signals from electrodes stuck on the scalp (showing brain activity), around the eyes (showing eye movements), and on the chin (showing muscle activity). The electrical signals were amplified to make them strong enough to flick tiny pens back and forth, leaving continuous squiggles on a strip of paper that scrolled underneath them. (Today, the electrical signals are recorded digitally, which enables many different graphic representations and data analyses.)

Aserinsky and Kleitman decided to leave the paper running all night. What they saw were recurring bouts of rapid eye movements that appeared roughly every 90–100 minutes across the night. (If you watch a sleeping child, cat or dog you can sometimes see their eyes flitting around under their closed eyelids.) Aserinsky and Kleitman also discovered that during these rapid eye movements, people's heart rate and breathing got faster. If you woke them up, they consistently recalled dreams. In contrast, if you woke them up when there were no rapid eye movements, they did not recall dreaming.

William (Bill) Dement, a giant among the founders of current sleep science and medicine, was a second-year medical student working with Aserinsky and Kleitman. He is credited with coining the term 'rapid eye movement sleep' (REM) for this phenomenon and describing the rest of sleep as 'non-rapid eye movement sleep' (NREM). Bill died at the age of 91 during the writing of this book, and his professional and personal legacy cannot be overstated.

Building on this seminal work, subsequent studies at the University of Chicago and in Lyon confirmed that there are indeed two distinct states of sleep, and further defined their differences as detected by polysomnography.

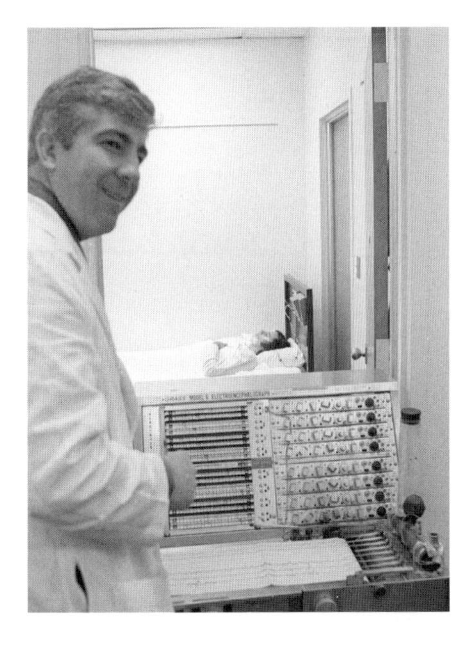

Figure 3.1 William Dement with a paper-based polysomnography system, 1970.[2] Photo: Jose Mercado, Stanford Medicine

An older story

In 1995, Kleitman gave a memorable presentation at his hundredth birthday celebration, during the annual meeting of the Associated Professional Sleep Societies in Nashville. I recall a little old man going up to the podium with pages of large-print notes, which he didn't look at during his speech. Among other things, he spoke about the discovery of REM sleep, but what has stayed with me most was his enduring passion and enthusiasm for research into the mysteries of sleep.

In a session later that day, I was chatting with an older physician from India who suggested that I should read the Upanishads, the texts at the spiritual core of Hinduism that are thought to be over 2500 years old. In these beautiful texts (in English translation),[3] I found multiple references to dreaming sleep and deep non-dreaming sleep, as in the following examples. (In each, 'the Self' refers to pure consciousness, the Brahman.)

In the Aitareya Upanishad:

> The Self being unknown, all three states of the soul are but dreaming – waking, dreaming, and dreamless sleep. In each of these dwells the Self: the eye is His dwelling place while we wake, the mind is His dwelling place while we dream, the lotus of the heart is His dwelling place while we sleep the dreamless sleep.

In the Kaivalya Upanishad:

> He, as the Self, resides in all forms, but is veiled by ignorance. When He is in the state of dream that men call waking, he becomes the individual self and enjoys food, drink and many other pleasures. When he is in the state of dream that men call dreaming, he is happy or miserable because of the creations of his mind. And when he is in the state of mind that men call dreamless sleep, he is overcome by darkness, he experiences nothing, he enjoys rest.

Figure 3.2 Vishnu dreaming the universe into existence: relief panel in the temple of Vishnu, Deogarh, Uttar Pradesh, India, *c.* 450–500 BCE. Photo: Arnold Betten

Figure 3.3 Painting, possibly a man in REM sleep, in Lascaux Caves, Dordogne, France, *c*. 17,000 BCE. Photo: iStock, https://www.istockphoto.com/photo/the-shaft-of-the-dead-man-cave-painting-in-lascaux-cave-france-gm1205277706-347136671

The pioneering French sleep scientist Michel Jouvet speculated, albeit with reservations, that a painting (Figure 3.3) in the Lascaux Caves (dated around 17,000 BCE) might indicate that some early modern humans (Cro-Magnons) understood another physiological aspect of dreaming.[4] Erections (clitoral tumescence in women) occur regularly during REM sleep, irrespective of dream content. Jouvet identified four elements in this painting and proposed that the artist wanted to simultaneously represent the dreamer, the concept of dreaming, and the content of the dream. There is the man lying on his back with an erection, indicating that he is dreaming. There is a bird mounted on a staff, which could represent the soul of the man leaving his body while he is dreaming (like Ba in ancient Egyptian mythology). Next to the man is an injured bison whose intestines are spilling from its stomach, and beside it a broken spear. These could represent what the man was dreaming about.[5]

These examples stay with me as a reminder that knowledge based on repeatable 'objective' measurements in the outside world – the sort of knowledge that I contribute to as a quantitative scientist – does not capture the full extent of human understanding and experience. This is also clear in research on the content and functions of dreaming, where essential data come from the subjective dream recall of participants.

Dreams as data

For much of recorded history, dreams were thought to contain important, even prophetic messages that could not be accessed any other way, and they were often attributed to an external source. Until the identification of REM sleep and its association with dreaming, the only source of information about dreaming was people's waking recall of what they had 'experienced'.

Most of us probably dream regularly, since people whose sleep is monitored using polysomnography typically have REM sleep.[6] However, the brain systems responsible for recent memory are turned off during REM, so dreams are seldom recalled unless you wake up from them. Thus, dreams remembered after spontaneous waking are only a small sample of our nightly ventures into this other form of consciousness. Nevertheless, dream journals remain an important source of data for dream research.

Another way of gleaning information about dreaming is to use polysomnography to monitor people's sleep in the laboratory and systematically wake them up from REM sleep to report on what they remember. Dream recall is more consistent in this setting, but there are also some differences in reports gathered in this way, compared to dream recall after spontaneous awakenings. Not surprisingly, people woken from REM sleep in the laboratory often report thoughts, feelings, and precepts relating to the laboratory situation. However, they still report the hallucinations, delusions, and bizarreness that characterise dreams recalled after spontaneous awakenings. The emotional content also seems to be more positive overall in dreams reported when people are woken from REM sleep in the laboratory, compared to reports after spontaneous awakenings at home.

A method that overcomes some of these challenges is to monitor participants' sleep at home so that the sleep state that they wake from spontaneously (REM or NREM) can be identified later and matched with their reports of what they recall experiencing immediately before each awakening. Participants can also be beeped when they are awake and asked to report on their waking consciousness.

Harvard psychiatrist Allan Hobson, one of the most influential modern dream researchers, has argued that for dreaming to be studied scientifically, we must shift away from the traditional focus on trying to interpret the content of dreams. Instead of asking 'what does a dream mean?', we need to ask 'what are the mental characteristics of dreaming that distinguish it from waking?'. This puts the focus on the formal properties of dreams – how we perceive things (perception), how we think (cognition), and how we feel (emotion) in dreams. In this paradigm, dreaming is a mental experience associated with REM sleep and its characteristic activation in some areas of the brain, deactivation in others, and changes in the connectivity among different brain areas.[7] There are other forms of mental activation during NREM and when falling asleep, but they do not have the same formal properties as REM dreams.

Psychologist Roslyn Cartwright earned the nickname 'the Queen of Dreams'. In her superb 2010 book *The Twenty-Four-Hour Mind*, she argued that it is time to put together what has been learned about the sleeping mind with what psychologists know about waking cognitive and emotional behaviours.[8] She provided a beautiful summary of the two parts of this circadian rhythm in our emotional lives:

Figure 3.4 Henri Rousseau, *The Sleeping Gypsy*, 1897. Museum of Modern Art, New York

> All day the conscious mind goes about its work planning, remembering, and choosing, or just keeping the shop running as usual. On balance, we humans are more action oriented by day. We stay busy doing, but in the inaction of sleep we turn inward to review and evaluate the implications of our day, and the input of those new perceptions, learnings and – most important – emotions about what we have experienced.

Cartwright died at the age of 98 during the writing of this book. She is acknowledged as a pioneer in the study of the links between dreaming and REM sleep and a trailblazer for women in sleep science. Her view of the role of dreams in the 24-hour mind was that they serve to regulate emotion and update the self.

At the time of this writing, an overview in the latest edition of the major textbook *Principles and Practice of Sleep Medicine* concluded that research on dreaming is still 'a field that is very much a work in progress'.[9]

Ongoing technological change and new knowledge

Technological advances continue to enable major steps forward in the modern scientific understanding of sleep. Polysomnography now is digitised, as noted earlier, and sometimes the identification of sleep states and stages is done by a computer algorithm rather than by a human technician. Nevertheless, the final interpretation of all that data is still usually entrusted to a human, notably for the diagnosis of sleep disorders. However, polysomnography, for all its advantages, is sometimes described as 'listening to crowd talk on the surface'. It doesn't give you a detailed picture of what is going on in the brain, or where things are happening, unless you add a lot more electrodes and complex computer algorithms to spatially interpret the array of electrical signals.

Particularly in the last two decades, major technological advances in modern sleep science have come with the advent of functional brain imaging technologies, notably positron emission tomography (PET), functional magnetic resonance imaging (fMRI), and single photon emission computed tomography (SPECT). These techniques can show the dynamic patterns of activity in different parts of the brain and how they change in different stages of sleep and in levels of alertness during wake. These four-dimensional (including time) visualisations of what is happening in the brain continue to amaze me.

Brain imaging technologies and polysomnography provide detailed objective data, but they are invasive, expensive, and not always practical. For example, they are not (yet) feasible for continuously monitoring sleep/wake patterns over many days. As we saw in Chapter Two, one way of doing this is to wear an actigraph on the non-dominant wrist. The data it records – usually every minute – are downloaded to a computer and fed though an algorithm that decides for each minute whether the wearer was asleep or awake.

Importantly, the algorithm has previously been validated by the manufacturer in experiments comparing simultaneous recordings of actigraphy and polysomnography. Actigraphy provides extremely useful information about people's sleep/wake patterns, for example during shift work or long-haul flights. Actigraphy can also be used clinically, for example to document the sleep/wake patterns of people who have trouble staying synchronised to the 24-hour day/night cycle.

What is the normal pattern of adult sleep?

Normal sleep for adults is widely believed to be eight hours of unbroken sleep at night. This is currently being debated, based on findings from three very different types of research: 1) a review of descriptions of sleep in historical Western European literature and other written sources;[10] 2) experimental studies with 14 versus 8 hours of darkness;[11] and 3) a study that used actigraphy to monitor the sleep of three hunter-gatherer societies in Tanzania, Bolivia, and Namibia.[12] Comparing these studies is an interesting exercise.

In his 2001 paper, historian Roger Ekirch argued that there has been an 'ingrained historical indifference' towards the timing and length of sleep and how it was influenced by societal position.[13] In response, he undertook an extensive review of diaries, medical books, imaginative literature, and legal depositions, looking for historical observations about sleep habits and patterns. He focused primarily on pre-industrial England with some additional reference material from Western European countries. This led him to conclude that prior to the industrial revolution, most people had two sleep periods per night, bridged by an hour or more of quiet wakefulness. The few exceptions come from documents by very wealthy people who could afford greater access to artificial light. One piece of supporting evidence cited medical textbooks from the fifteenth to the eighteenth centuries, which frequently advised sleepers to improve their digestion and get more tranquil sleep by lying on their right side during the first sleep and on their left side before the second. Ekirch notes that split sleep at night is recorded as far back as Homer's Odyssey, written around the end of the eighth century BCE.

The commonalty of references to waking spontaneously for a period in the middle of the night, and the fact that this attracted minimal additional comment, reinforced Ekirch in his view that this was the norm. The period of wakefulness in the middle of the night was variously considered to be a time for prayer, contemplation, sex, household chores, pressing work demands, or assorted nefarious activities. Another common feature in Ekirch's sources of information was reference to dream recall upon awakening from the first sleep, and the value of having undisturbed time to contemplate the content of dreams during the ensuing few hours of wakefulness before the second sleep. He laments:

> It is no small irony that, by turning night into day, modern technology, while capable of exploring the inner sanctums of the brain, has also helped obstruct our oldest avenue to the human psyche.[14]

He maintains that the availability of artificial lighting is the key difference between cultures where split sleep is, or was, the norm and modern industrialised societies. With the changes in work patterns that accompanied the industrial revolution and the increasing availability of artificial light, the second half of the nineteenth century saw active public promotion of early rising – people were encouraged to gain time and 'embrace modernity' by giving up the lethargy and lust of the second sleep. Only a small section of the middle class opposed this as 'unnatural'.

The action of artificial light in determining one versus two sleeps per night was demonstrated in the 1990s in a series of experiments by Tom Wehr and colleagues at the US National Institute of Mental

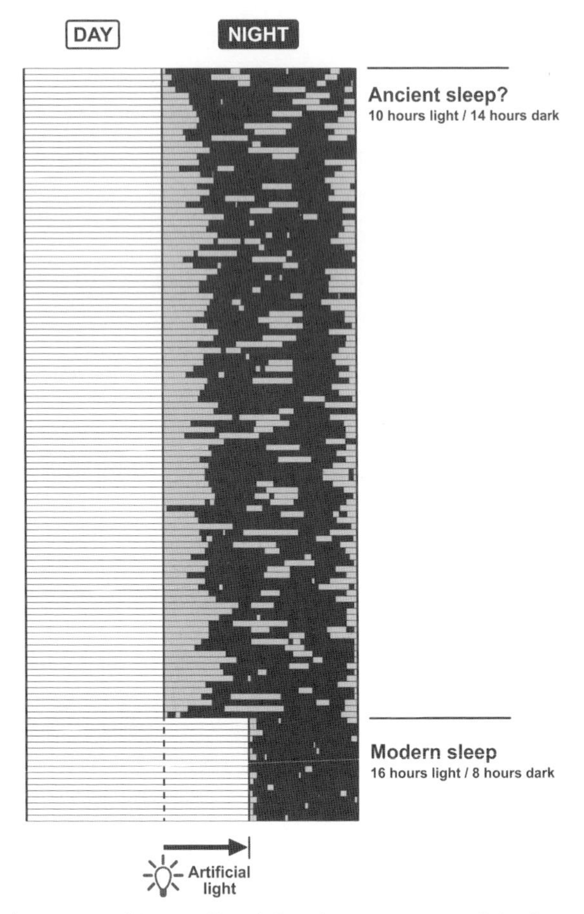

Figure 3.5 Polysomnographic recording of sleep in a young woman living in two different lighting regimens: 14 hours of darkness per 24 hours (upper panel), and eight hours of darkness (lower panel). Wehr, T. A. (2001). Photoperiodism in Humans and Other Primates: Evidence and Implications, *Journal of Biological Rhythms 16*(4), 348–364

Health. Figure 3.5 shows the sleep of a young woman living in two different light regimens: the first with 14 hours of darkness per 24-hour period, the second with eight hours of darkness.[15] In both regimens, during each daily light period, she went about her normal activities and so was exposed to the ambient artificial and natural light in her normal environment. During each daily dark period, she was required to be in a windowless, dark room and instructed to remain in bed (except to use an adjoining dark bathroom) and to sleep whenever possible. Each strip across the figure represents one day. The black bars show when polysomnography indicated that the young woman was asleep.

In the upper part of the figure, the woman lived in a cycle of 10 hours of light followed by 14 hours of darkness (LD 10:14) for more than 15 weeks. Her sleep expanded and separated into a typical pattern of two bouts of three to five hours with a one- to two-hour period of wakefulness in between. The average total sleep of 15 healthy young people exposed to this regimen was significantly longer (10.6 hours) than when they lived in the regimen with only eight hours of darkness (7.6 hours). When the lighting regimen changed to 16 hours of light followed by eight hours of darkness (LD 16:8), sleep was compressed and consolidated.

Another piece of intriguing evidence supporting the argument that adult sleep is naturally split came from a survey of nearly 2000 people who had been prescribed sleeping pills to help them get to sleep at night. It found that 20 per cent were not using the pills at the beginning of the night, they were using them to get back to sleep in the middle of the night.[16]

In contrast, in a 2015 paper, anthropologist Ghandi Yetish and a distinguished team of US-based anthropologists and sleep researchers dispute the idea that split sleep is normal for adult humans. They base this on actigraphy recordings of the sleep patterns of adults (47 females and 47 males) in three current hunter-gatherer societies, the Hazda in Tanzania (2 degrees south), the Tsimane in Bolivia (15 degrees south), and the San in Namibia (20 degrees south). Among these peoples, consolidated sleep at night was the norm, with sleep beginning on average 3.3 hours after sunset and wake-up usually before sunrise. Napping occurred on less than 7 per cent of days in winter and less than 22 per cent of days in summer. The average sleep duration identified by actigraphy was 5.7–7.1 hours (average time between initial sleep onset and final wake-up

6.9–8.5 hours). Sleep averaged one hour longer in winter than in summer, mainly due to variation in the time of going to sleep. In these equatorial climates, light exposure was greatest in the morning, with people typically seeking shelter from the heat of the sun in the middle of the day.

Yetish and colleagues argued against Ekirch's hypothesis, stating that the disappearance of two sleep periods was not 'a pathological development caused by reduced sleep duration', but rather a return to an ancestral sleep pattern enabled by the control of lighting and temperature in industrialised societies which 'restored aspects of natural conditions in the tropical latitudes'.[17]

In an attempt to see whether the findings from these three studies could be integrated in a broader framework, I decided to look at the seasonal changes in day length that occur at the latitudes in which each of the study populations lived. As Figure 3.6 illustrates, each study drew participants from different latitudinal bands. The figure assumes that participants in the studies of Wehr and colleagues came from across the continental USA (this was not reported). It is possible they lived in a much narrower latitudinal range, but it would still be expected to be intermediate between those whose written records informed Ekirch's hypothesis and the hunter-gatherers monitored by Yetish and colleagues.

Split sleep patterns were found in Wehr's experiments when people were exposed to 14 hours of darkness per 24 hours, but not when they were exposed to eight hours of darkness per 24 hours. Figure 3.6 suggests that people living in England at 50 degrees north (the solid grey curve) would experience at least 14 hours of natural darkness (less than 10 hours of daylight) for about four months of the year. In contrast, the hunter-gatherers living between 2 and 20 degrees south would never experience 14 hours of natural darkness.

These comparisons highlight both the strengths and weaknesses of different research approaches, and the value of trying to integrate their findings.

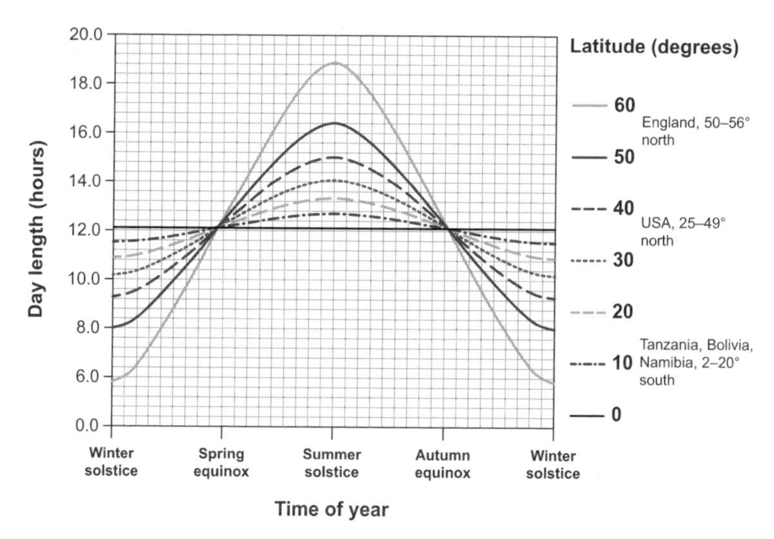

Figure 3.6 Diagram of daylength changes across the year, at different latitudes. The solid black horizontal line indicates that daylength at the equator is around 12 hours irrespective of the season. By comparison, the solid light grey curve indicates that daylength at latitude 60° (north or south) ranges from around six hours at the winter solstice to around 19 hours at the summer solstice.

Lessons from arts–science collaboration

I have learned a great deal about the limits of my own research in my arts–science collaborations with Sam Trubridge, a very insightful performance designer and director. Sam has a knack for challenging me about things that I take for granted and then turning these into artistic projects that challenge others as well.

For instance, I explained to him how we use the actigraphy data collected in our research with airline pilots. We monitor them before, during, and after international trips, then de-identify all the data so it can't be traced back to any individual pilot, and combine everyone's records for statistical analysis. From an ethical and quantitative science perspective, this is currently considered best practice.

Sam pointed out that it means we remove the personal narrative that belongs to each actigraphy recording and lose a great deal of relevant information in the process. After some vigorous discussion, we agreed that we would each wear an actigraph for six weeks and photograph our beds when we got up every morning. This would reinsert a token reminder of what else was happening in our lives at the time, in addition to when we moved our non-dominant wrists.

Figure 3.7 shows nine days of Sam's actigraphy record. Each horizontal panel represents 24 hours, from noon to noon, with the photos showing the state of his bed when he woke up each morning. The shaded area each day covers the 16 hours from 6 am to 8 pm (New Zealand Standard Time). Each vertical black bar represents the amount of wrist activity in one minute. Where there are long periods with very little movement, Sam was asleep.

Sam tended to go to bed very late on Friday night and wake up late on Saturday and Sunday mornings. During the week, he had to get up earlier to go to his job as a university lecturer. He regularly got less than seven hours sleep when he had to go to work the next morning. This pattern suggests that Sam has a fairly extreme evening chronotype. The photos suggest some nights were more restless than others, but I don't know the details.

During my six weeks of recording, I travelled from my home in Wellington to Notwill, Switzerland, to be part of a team convened by the World Health Organization to look at the impact of sleep disorders on people's lives. There are two photos of my aeroplane seats associated with long periods with very little sleep, and there are differences in how disturbed my bedding was on different mornings in both places, but I no longer remember the details. Nevertheless, the photos serve as a reminder that the diverse experiences of our daily lives were happening during these recordings. Sam made our actigraphy records and photos into two massive banners for an installation work named *On My Side*.

From this chapter, it is clear that quantitative science does not capture all the richness of our experience of sleep, and that other

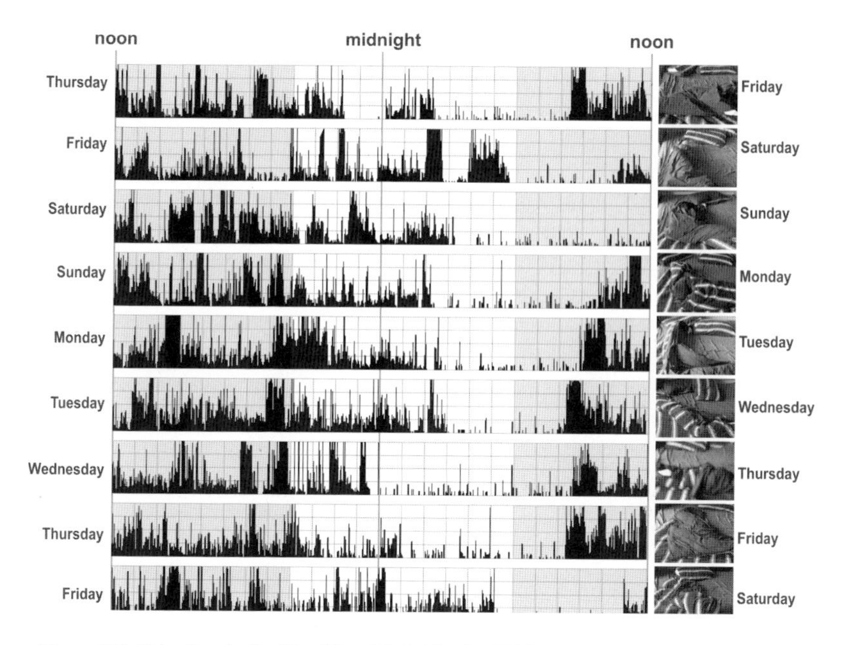

Figure 3.7 Nine days in the life of Sam Trubridge in 2009. Photo: Sam Trubridge

types of knowledge can add to our understanding. However, quantitative science does provide compelling evidence debunking the myth that we can get more out of life by sacrificing sleep. This is the focus of Chapter Four.

Chapter Four

Sleep – Why We Need It

As the Upanishads recognise, we live in (at least) three different brain states – waking, dreaming, and dreamless sleep. As a chronobiologist, I view these as integrated components of our continuous circadian rhythms. It is reasonable to expect that essential things are happening during sleep if we are programmed to spend a third of our lives in this complex series of 'off-line' states. It is also reasonable to expect that what happens when we are asleep influences how we function when we are awake and vice versa. We each have the same brain and body 24 hours a day.

I first delved into the mysteries of sleep through a serendipitous affiliation with the legendary Stanford University Sleep Disorders Center led by Bill Dement. At NASA in the 1980s I was considered an 'alien' – the US government term for a foreign national. I must admit that I enjoyed the idea of being an alien at NASA, but a consequence was that I was not able to be employed directly by them. I was contracted through other organisations, including a year as a consultant at the Stanford Center. Although I was based down the road at NASA Ames, this enabled me to learn from and work with some truly exceptional sleep scientists and clinicians.

I am convinced that the more we learn about sleep and its functions, the more evident it becomes that we cannot trade it for more waking hours. This chapter reviews the evidence.

What happens during normal sleep

As described in Chapter Three, polysomnography has identified the two different states of sleep described as rapid eye movement (REM) sleep and non-REM (NREM) sleep, which occur in mammals and birds (at least). These two states alternate in a regular cycle across normal sleep at night and are as different from one another as each is from wake. Three types of electrical activity are monitored to identify REM and NREM.

Electrical activity in the brain is monitored by electrodes on the forehead and scalp, with ground electrodes behind each ear. Eye movements are monitored by electrodes beside each eye. Muscle tone is monitored by electrodes on either side of the chin.

Figure 4.1 Sleep scientist Margo van den Berg wearing basic polysomnography electrodes. Photo: Massey University

NREM sleep is divided into three stages based on progressive changes in the brain's electrical activity: light sleep (Stages 1 and 2) and deep slow-wave sleep (Stage 3).

We usually fall asleep into Stage 1 NREM. The exact moment of falling asleep can be difficult to pinpoint. In general, as sleep approaches muscle tone decreases (muscles relax) and slow-rolling eye movements begin to occur, although a person may still say that they are awake when this is happening. The onset of Stage 1 NREM is signalled by a change in the electrical patterns in brain activity. During quiet wakefulness, there is a clear rhythmic pattern of brainwaves in the range of 8–13 cycles per second (known as alpha activity). In Stage 1 NREM, this pattern is replaced by mixed-frequency, low-amplitude brainwaves. Even with this change in brain activity, people sometimes say they are awake. Stage 2 NREM is marked by the appearance of signature bursts of synchronised brain activity known as 'K complexes' and 'sleep spindles'. This is accepted by most sleep researchers and clinicians as an unequivocal sign that a person is asleep.

Interestingly, after being asleep for about 10 minutes, we don't remember things that happened during the last few minutes of wake. This explains why a snooze alarm that recurs at intervals shorter than 10 minutes gives you a better chance of remembering that it has already gone off.

As Stage 2 NREM progresses, slow-frequency (0.5 to 4.5 cycles per second), high-amplitude brainwaves occur more often. Like a crowd at a football match, specific areas of the brain progressively join a big 'wave' where their neurons all fire at the same time. Stage 3 NREM occurs when these 'slow waves' account for at least 20 per cent of the total measured brain activity.

As we progress deeper into sleep from Stage 1 to Stage 3 NREM, muscles become more relaxed, and heart rate and breathing slow down. The brain is also disengaging progressively from stimuli coming from the outside world, making it harder and harder to wake the sleeper up. Meaningful stimuli, such as hearing your own name, or a mother hearing her own baby cry, are more likely to get through and wake you up. NREM is sometimes described as an inactive brain (although it is still regulating vital functions) in a moveable body (i.e., a body that still receives and responds to signals from the brain instructing it to move).

In contrast, during REM sleep, from time to time the eyes dart around under closed eyelids and this is often accompanied by muscle twitches and fluctuations in heart rate and breathing. The electrical patterns in brain activity during REM are desynchronised (spread across a range of different frequencies) and look much like waking brain activity. About 80 per cent of the time, people woken from REM have vivid dream recall. However, during REM sleep we are effectively paralysed. A block in the brainstem stops brain signals reaching the muscles involved in movement, which show minimal electrical activity. This normally protects us from acting out our dreams. Occasionally people experience brief paralysis if they wake up just before the brainstem block switches off. REM sleep is sometimes described as an active brain in a paralysed body.

In young adults during a normal night of sleep, the first bout of REM usually occurs 80–100 minutes after falling asleep, and thereafter REM bouts recur roughly every 90 minutes. Most deep, slow-wave sleep (NREM Stage 3) occurs in the first few NREM/REM cycles of the night. Towards morning, REM bouts get longer and more intense and mostly alternate with light NREM (Stages 1 and 2). Generally, we only remember dreams if we wake from them. During the week, I set the alarm so I can get to work on time. On Saturday I like to give myself the treat of waking spontaneously, and sometimes am rewarded with a fleeting recollection of a dream, something I rarely experience during the week.

There are marked age-related changes in the internal structure of sleep (also known as 'sleep architecture'). Newborn babies typically fall asleep into active sleep (the precursor of REM) before quiet sleep (the precursor of NREM). They also have a shorter sleep cycle of about 50 minutes and spend about half of this in active (REM) sleep. Over the first two years, the amount of REM declines to about 20–25 per cent of sleep and the slow waves characteristic of NREM Stage 3 emerge. Across adolescence, NREM Stage 3 decreases by about 40 per cent and continues a slower decline into old age, particularly in men. In contrast, REM takes up about 20–25 per cent of sleep from childhood through adolescence, adulthood and into old age, except in dementia.

Why two kinds of sleep?

When sleep is too short, the amount of time spent in slow-wave sleep tends to be conserved at the expense of REM sleep. In recovery sleep after sleep deprivation, the lost hours are not recovered hour-for-hour. Instead, there is extra slow-wave sleep on the first recovery night and sleep may be slightly longer than usual. In fact, if the pressure for slow-wave sleep is very high on the first recovery night, it can push REM recovery out to the second night. After two nights of unrestricted recovery sleep, the NREM/REM cycle usually returns to normal.

Why is deep slow-wave sleep so important that it gets priority in short sleep and recovery sleep? As described earlier, slow waves are caused by synchronised cycles of firing and resting involving a very large number of neurons in the neocortex area of the brain. They are associated with consolidation of some types of memory, and a recent brain imaging study suggests they have another vital function.[1] About every 20 seconds during slow-wave sleep, while the neurons are 'resting' after a peak in slow-wave activity, blood pressure in the brain drops and cerebrospinal fluid (CSF) flows into fluid-filled spaces in the brain (the ventricles). As the slow-wave activity builds again (the neurons start firing again), the CSF is pumped out, carrying away metabolic wastes produced by brain activity. In older adulthood, the amount of slow-wave sleep declines. It is plausible that the resulting reduction in CSF cleansing of the brain might contribute to impaired memory with aging, and to dementia.

On the other hand, REM sleep seems to be involved in the consolidation of recently acquired emotional memories and may influence emotional reactions during subsequent wake. It is plausible that remembering salient emotional experiences might improve your chances of survival in some situations.[2] The large amount of active sleep in newborn babies and the decline in REM sleep across infancy suggest that it also has a role in brain development, and it may be specifically involved in learning new motor skills. A possible mechanism for how REM might affect brain development, learning, and memory consolidation has also been proposed. There is evidence that during REM sleep, existing connections between some neurons are pruned and new connections are reinforced.[3]

New evidence is accumulating rapidly about the multiple different functions of NREM and REM sleep. On reflection, it is not surprising that these two sleeping brain states serve multiple functions, as does waking.

Sleep timing

Two main processes interact to regulate the timing and structure of sleep: 1) the circadian master clock in the SCN and 2) a pressure for sleep that builds up in the brain across wakefulness and gets discharged during sleep.

The circadian master clock influences when REM sleep occurs. It drives a circadian rhythm in REM propensity, which peaks just after the circadian low point in temperature (typically around 3–5 am during a normal night's sleep). If you fall asleep around this time, you will go into REM more quickly, and this is also when REM bouts are longest and most intense.

The circadian master clock also influences when we fall asleep and when we wake, by acting on competing sleep-promoting and wake-promoting brain areas. The solid curve in Figure 4.2 visualises this in terms of a circadian rhythm in the drive for the brain to be awake.

The 'circadian wake drive' is at its weakest around 3–5 am, when sleepiness peaks and we are also around our least functional and most error prone (the exact time is earlier in morning-types than evening-types). The wake drive then builds up again, towards a brief plateau in the afternoon – the 'nap window'. After the nap window, it continues to rise into the mid-late evening (again the timing varies by chronotype). In the few hours before normal sleep time, when the wake drive is at its strongest, it is very difficult to fall asleep. This is known as the 'evening wake maintenance zone'. It explains why we find it difficult to fall asleep a few hours earlier than usual in anticipation of an early wake-up the following morning.

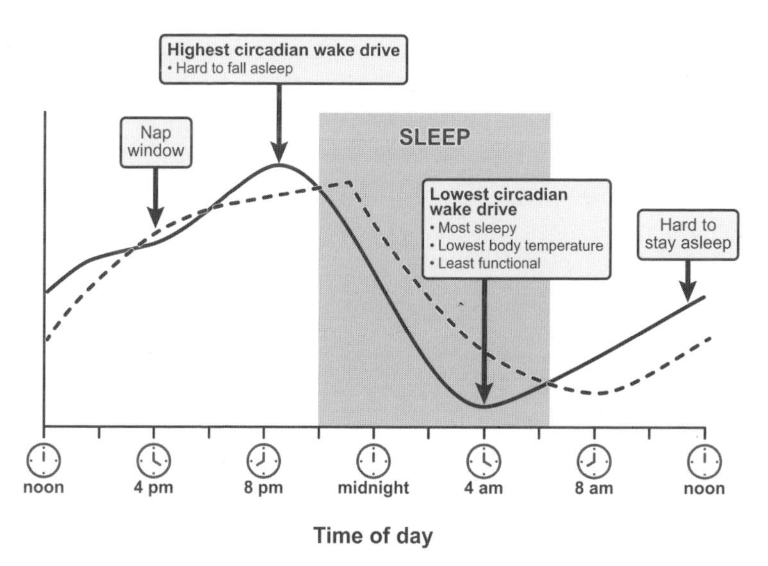

Figure 4.2 A visualisation of how the circadian wake drive (solid curve) interacts with the rise of homeostatic sleep pressure across wake and its exponential decline across sleep (dashed curve). Wake is favoured when the solid curve is above the dashed curve. Sleep is favoured when the dashed curve is above the solid curve.

The second process affecting sleep timing is the pressure for sleep that builds up in the brain across wakefulness and gets discharged during sleep. This is known as the homeostatic sleep drive (the dashed curve in Figure 4.2). If you stay awake longer than usual, and particularly if you haven't had enough sleep recently, you will feel the increasing power of this pressure for sleep. We often try to counteract it, for example by taking a caffeine hit with coffee or an energy drink, but the effects of stimulants are only temporary. The only way to reduce the pressure for sleep is by sleeping.

Homeostasis refers to the way in which many processes in the body are kept within certain preset limits for optimal functioning. For example, when your blood sugar levels rise, your pancreas produces insulin to bring them down again. Similarly, as the length of wakefulness increases, the brain responds by producing pressure for sleep. The homeostatic pressure for sleep is reflected in the amount of brain activity that is in the slow-wave frequency range (0.5–4.5 cycles per second). If you didn't get enough sleep last night or have been awake longer than about 16 hours, extra homeostatic pressure for sleep builds up and this is reflected in more slow-wave activity than usual in the first couple of NREM/REM cycles of recovery sleep.

It is not yet clear what causes the pressure in the brain for sleep, but one possibility is the build-up of adenosine during wakefulness and its subsequent decline during sleep. Adenosine is a by-product of energy usage by brain cells and it depresses brain activity. During time awake, the amount of adenosine increases, particularly in the areas of the brain that are actively promoting wakefulness, so these areas become increasingly inhibited. During sleep, brain activity drops and so do adenosine levels. Caffeine stimulates the brain by blocking the action of adenosine on brain cells. This may explain how caffeine can temporarily block the homeostatic pressure for sleep.

The interaction between the circadian pacemaker and homeostatic sleep drive creates 'windows' when sleep is easier or when sleep is more difficult. After its low point around 3–5 am, the circadian wake drive starts to strengthen. At the same time, the homeostatic pressure for sleep is reaching its lowest level because it has been dissipating across sleep. After the wake drive has been rising for about three hours, it overwhelms the low homeostatic pressure for sleep and wake-up occurs. As the wake drive continues to increase across the morning it becomes increasingly difficult to stay asleep. This often truncates sleep after night work.

The afternoon plateau in circadian wake drive provides a second window for sleep, particularly if the homeostatic pressure for sleep is higher than usual because you haven't had enough sleep recently. This is another time when you have increased risk of falling asleep at the wheel of a motor vehicle (although the greatest risk is when the circadian wake drive is at its lowest).

After the nap window the circadian wake drive increases again, and so does the homeostatic pressure for sleep. Once the circadian wake drive passes its peak, the homeostatic pressure for sleep overwhelms it and the night's sleep begins. This is typically about five hours before the circadian wake drive gets back down to its 3–5 am minimum. Again, morning-types usually fall asleep earlier than evening-types.

Sleep debt

The pressure for sleep builds up not only across hours awake, but also across multiple days when sleep is too short or of poor quality. This is known as building up a sleep debt.

Sleepiness is a sign that we need sleep, but it is often misconstrued as laziness, lack of interest, or poor motivation (whereas we generally accept that hunger and thirst indicate that we need to eat and drink). As sleepiness increases, many other aspects of waking function are deteriorating, leading to increasing irritability, degraded alertness, slower reaction times, poorer coordination, slower thinking, loss of situation awareness, and less creative problem-solving.

Sleepiness can be insidious because as the pressure for sleep builds, we become less reliable at judging how well we are functioning. Laboratory studies indicate that after the first two or three nights of sleep restriction, people feel increasingly sleepy. However, with additional nights of sleep restriction, they report feeling no sleepier, even though on objective measures their performance is continuing to deteriorate. This is relevant to the idea of social jet lag that was introduced in Chapter Two. A person who works Monday to Friday and must use the alarm clock to wake up two hours earlier than their spontaneous wake-up time in order to get to work on time, will have accumulated 10 hours of sleep debt before their commute on Friday morning. On Friday they will be less functional than on Monday and may well be unaware of it. The effects of sleep restriction are also dose-dependent – the more sleep is cut short each night, the faster impairment builds.

The pressure for sleep continues to build to a point where people fall asleep uncontrollably, even in life-threatening situations. These microsleeps may only last seconds, but a vehicle travelling at 100 kilometres per hour travels 278 metres in 10 seconds. Well before a driver reaches the point of falling asleep uncontrollably, their lane position and speed control become more erratic and their risk assessment is deteriorating. I have provided expert testimony in too many investigations of fatal crashes caused by sleepy drivers.

How do we pay back a sleep debt and return to optimal waking function? Currently science cannot provide a clear answer to this extremely important practical question. As noted earlier, after two nights of unrestricted recovery sleep, the NREM/REM cycle usually returns to normal. However, laboratory studies that have looked at recovery after multiple nights of restricted sleep suggest that full recovery of waking function may take longer than two nights of unrestricted sleep.[4] After seven nights of sleep restriction in the laboratory, participants who were allowed two 10-hour recovery sleep opportunities at night or three eight-hour opportunities at night did not show full recovery of waking function. There are also differences in recovery rates on different performance tasks, and some people are consistently less affected by sleep loss than others. Given these uncertainties, from a practical standpoint the current recommendation is that at least two consecutive nights of unrestricted sleep are needed for recovery from sleep debt. Chapter Five introduces some ideas about how to reduce sleep debt related to work patterns, and how to manage its consequences.

Sleep is definitely not an optional 'off time', although that myth is surprisingly enduring.

Figure 4.3 *Sleep/Wake*, Auckland Festival 2009. Director/designer, Sam Trubridge; sleep scientist, Philippa Gander. The Orator (Jamie Burgess) introduces the audience to the intricacies of NREM and REM sleep on scrolling polysomnography, before guiding them on a journey through a series of awakenings as the stage opens out and they become surrounded by the action. Photo: Richard Robinson

Why sleep?

Much of the evidence relating to why we sleep comes from studies that examine what happens when people don't get adequate sleep. This can be done by manipulating sleep experimentally in laboratory studies or by comparing people in large, population-based studies that either use one-off surveys or track cohorts of people across time. In 2016, the UK Royal Society for Public Health published the review 'Waking Up to the Health Benefits of Sleep' which makes compelling reading.[5] It draws together findings from a large body of research in this summary table (reproduced with permission).

A comprehensive review of the science behind this table would take up at least one extra chapter in this book. The relationship between sleep and obesity is particularly noteworthy, given that obesity is a major public health issue in many countries. In

Table 4.1 Some of the Main Consequences of Poor Sleep

Physical	Mental	Behavioural	Performance
Metabolic abnormalities.	Depression.	Sleepiness.	Impaired attention and concentration.
Weight gain and obesity.	Psychiatric relapse.	Road traffic accidents.	Decreased memory.
Reduced immunity.	Mood fluctuation.	Falls and fractures.	Reduced multi-tasking.
Cancer.	Delirium.	Repeat prescribing.	Impaired decision making.
Cardiovascular disease and stroke.	Impulsivity.	Alcohol and drug dependency.	Reduced creativity.
Disorders of the hypothalamic-pituitary-adrenal.	Anger and frustration.	Increased sedative and stimulant use.	Reduced communication.
Thermoregulatory problems.	Higher risk of suicide.	Less likely to attend appointments.	Reduced socialisation.
Vulnerable seizure threshold.	Anxiety and hyperarousal.	Longer stay in hospital.	Less likely to be employed.
	Chronic fatigue.	Earlier admission to long-term care.	More likely to be on benefits.
	Increased pain.		

population-based studies of adults, short sleepers are more likely to become obese. Young children who don't get adequate sleep are more likely to become obese in later childhood and to be obese as adults. In laboratory studies, restricting the sleep of healthy young adults alters the balance of the appetite-regulating hormones leptin and ghrelin, increasing food intake and changing food preferences.

In the context of the Covid-19 epidemic, it is vital to recognise that having adequate sleep on a regular basis is essential for maintaining the effectiveness of the immune system. After only a short period of reduced sleep, people are more vulnerable to infection and their immune systems respond less effectively to vaccination. In one striking example, Tanja Lange and her colleagues from the University of Lübeck published a study in 2011 that compared the effects of hepatitis A vaccination on two groups of healthy young men. One group was allowed to sleep normally on the night after vaccination and the other group had to stay awake on the night after vaccination.[6] The participants received three hepatitis A vaccinations at 8 am on days 0, 8 and 16 of the study. The immune response was greater after each vaccination in the group allowed to sleep, and they still had greater immunity a year later, indicating that sleep on the night after each vaccination had improved 'immunological memory'. Interestingly, the level of immunity a year later was related to the amount of slow-wave sleep on the first night after each vaccination.

Disturbances in waking mental function that define psychiatric disorders are usually accompanied by disturbances in sleep. A prevailing view has been that the waking disorder causes the sleep problems, but mounting evidence suggests that this relationship is often bi-directional, i.e., each exacerbates the other. Treatments that target sleep can improve outcomes for patients with anxiety and traumatic disorders. Reports of insomnia or hypersomnia (too much sleep) are nearly universal among people with unipolar major depression, and recurrence or worsening of sleep symptoms can predict the relapse of depressive symptoms. In bipolar disorder, disruptions to both sleep and circadian rhythms play a role in causing waking symptoms and functional impairment. Schizophrenia is often associated with severe insomnia, which can predict relapse. Advancing knowledge about the 24-hour brain has led to the recommendation that optimal clinical management of psychiatric disorders should concurrently assess and manage sleep symptoms as well as waking symptoms.

The evidence is clear. Sleep is one of the three pillars of health, alongside diet and exercise. A useful analogy is that health is like a three-legged stool. You need all three legs to stay in balance.

Sleeping to stay healthy

Healthy sleep is defined not only by what happens during sleep, but also by how we function and feel when we are awake. In fact, waking symptoms (excessive sleepiness, difficulty getting to sleep, etc.) are usually the trigger for investigating whether sleep is problematic. From a chronobiological perspective, this makes complete sense – we each have the same brain and body 24 hours a day. Nevertheless, current Western medicine still focuses primarily on waking symptoms and what happens during wake.

Daniel Buysse, a psychiatrist and sleep specialist at the University of Pittsburgh,[7] has proposed a general model of sleep health that has five dimensions:

1. sleep duration – the total amount of sleep obtained per 24 hours;
2. sleep continuity or efficiency – the ease of falling asleep and returning to sleep;
3. timing – the placement of sleep within the 24-hour day (I would add the placement of sleep in an individual's circadian master clock rhythm);

4. alertness/sleepiness – the ability to maintain attentive wakefulness; and
5. satisfaction/quality – a personal assessment of 'good' or 'poor' sleep.

The US National Sleep Foundation convened an expert panel to review current knowledge and agree on recommendations for how much sleep is needed to stay healthy across the lifespan. They defined three categories at each stage of life: 'recommended'; 'possibly acceptable', which recognises that there is considerable variability among people at all ages; and 'not recommended'. Their findings are summarised in Figure 4.4.

Very short sleep is often driven by trying to fit everything into a busy life and the common, but mistaken, belief that sleep can be sacrificed to get more done. On the other hand, very long sleep is often associated with chronic illnesses.

In healthy people, sleep quality can be disturbed by a range of environmental factors (light, noise, temperature) and in response to major life events (changing jobs, birth of a new baby, living with

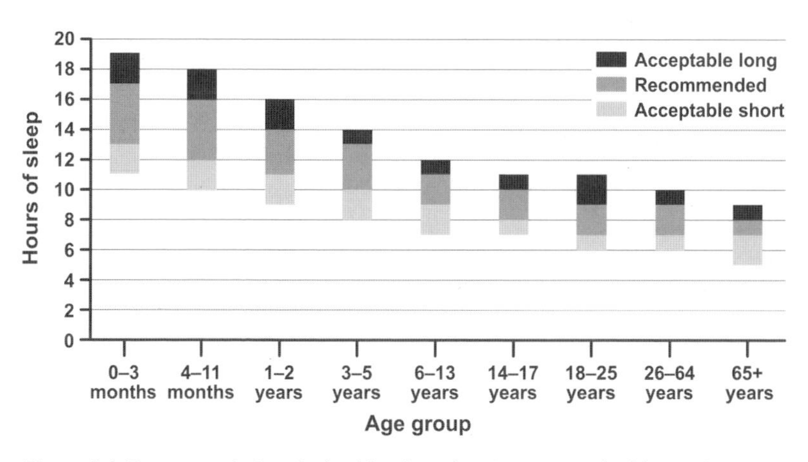

Figure 4.4 Recommendations for healthy sleep durations across the lifespan.[8]
Data From The Us National Sleep Foundation

teenagers, divorce, bereavement, etc.). On the other hand, people who have a sleep disorder are unable to obtain restorative sleep even under ideal conditions. Sleep disorders fall into six main categories.[9]

Insomnia refers to difficulties falling asleep or staying asleep. About 10–15 per cent of the adult population experience persistent insomnia, which has a negative impact on quality of life and bi-directional (cause and effect) relationship with poor mental and physical health.

Sleep-related breathing disorders are the result of intermittent collapse of the upper airway during sleep (obstructive sleep apnoea) and/or pauses in the respiratory mechanism that generates the breathing cycle (central sleep apnoea). Sleep apnoea is associated with excessive daytime sleepiness, increased risk of motor vehicle and work-related accidents, high blood pressure, elevated risk of strokes, impaired cognitive function, and reduced quality of life.

Narcolepsy is an autoimmune disorder which progressively destroys neurons that produce orexin, a critical wake-promoting hormone. It has four classic symptoms: sleep attacks (falling asleep uncontrollably); cataplexy (going limp when the REM movement block in the brainstem goes on during wake); falling rapidly into REM sleep with hallucinations at sleep onset; and being aware of sleep paralysis at the beginning and end of sleep.

Circadian rhythm sleep disorders arise when the SCN master clock is not synchronised to the 24-hour day/night cycle, or when it has a very long or short cycle length, so that people naturally sleep much later or earlier than societal norms (as discussed in Chapter Two). Chronic sleep disruption due to shift work or jet lag is also recognised as sleep disorder.

Sleep-related motor disorders are uncontrolled movements during sleep, such as periodic limb movements (usually leg twitches) or teeth grinding.

Parasomnias, or unusual behaviours during sleep, can occur during NREM, for example sleep walking and night terrors. They can also occur during REM sleep, for example recurrent nightmares and REM behaviour disorder (when the REM movement block in the brainstem fails and people act out their dreams).

For people with sleep disorders, the combination of being unable to get restorative sleep and not having (or choosing not to have) enough time to sleep can have severe consequences for their health, safety, and well-being. The availability of treatment services for sleep disorders varies widely between countries but is generally improving as their importance is increasingly recognised.

So how important is sleep? In its review, the UK Royal Society for Public Health concludes:

> Given its importance to our overall health and well-being, we would like to see a societal shift so that individuals are given the opportunity to get a healthy amount of sleep and offered support when they are having difficulties with sleep.[10]

Some commentators take this further and argue that sleep is a fundamental human right under the United Nations Universal Declaration of Human Rights, including the 'right to life' (Article 3); and the 'right to a standard of living adequate for the health and well-being' (Article 25).[11] For example, in a 2012 case of a night-time police raid on an encampment of sleeping protestors, Justice Chauhan of the Supreme Court of India wrote:

> An individual is entitled to sleep as comfortably and as freely as he breathes. Sleep is essential for a human being to maintain the delicate balance of health necessary for its very existence and survival. Sleep is, therefore, a fundamental and basic requirement without which the existence of life itself would be in peril.

For the financial year 2016–17, the estimated cost of inadequate sleep in Australia (population: 24.8 million) was $45.21 billion. The financial cost of $17.88 billion represented 1.55 per cent of gross domestic product, and the non-financial cost of $27.33 billion represented 4.6 per cent of the total Australian burden of disease for the year.[12]

In his book *The Promise of Sleep* (2000), Bill Dement labelled the introduction 'Prescription for Sleep-Sick Society'. He did not mince his words.

> For nearly half a century, a huge reservoir of knowledge about sleep, sleep deprivation, and sleep disorders has been building up behind a dam of pervasive lack of awareness and unresponsive bureaucracies. We don't know how many preventable tragedies are occurring right now, today, this very instant. It is time to either lower the floodgates or blow up the dam.[13]

He had given up on the authorities taking the former approach and was advocating the latter. His hope was that his book would be 'a bomb that blows up the dam and lets the information flow to the millions of people whose lives it can change – and save'. Progress has been made in the last two decades, but we are not there yet. In Chapter Five we consider the implications, for both our sleep and our waking lives, of our attempts to override our genetically programmed preference for sleep at night.

Chapter Five

Circadian Biology versus 24/7 Living

Circadian biology is an ancient legacy of life on our rotating planet. Trying to override it is a recent human choice. Chapter Two introduced some of the choices we now live with. Political decisions to redefine time zones in some countries mean that their inhabitants must live out of step with solar time (which moves one hour later for every 15 degrees of longitude that you travel east). Daylight saving moves the clocks (but not the day/night cycle) forward in spring and back again in autumn, to maximise people's opportunity to experience summer sunlight outside of what society dictates as standard work hours. Social jet lag arises because work hours do not allow some people adequate time to sleep during the optimum part of their circadian master clock rhythm on workdays. This results in sleep debt accruing across workdays, and unstable sleep times that switch back and forth on workdays versus days off.

This chapter takes a more detailed look at the two most major disruptions that we commonly impose on our circadian timekeeping system: shift work, and jet lag caused by flying across time zones. The most extreme example in terms of circadian disruption can be seen in airline pilots who fly international trips crossing time zones and simultaneously experience the effects of both shift work and jet lag.

In all these conflicts between solar time and social time, we compromise a vital aspect of our basic biology – the internal synchrony among our myriad circadian rhythms and their external synchrony with the day/night cycle. New scientific understanding has exposed these underlying biological challenges, and it also provides the basis for new strategies to reduce their negative impact on our health, safety, and well-being. The chapter ends with some examples.

When work patterns conflict with the day/night cycle and circadian time

Our circadian timekeeping system is an adaptive response to optimise our functioning across the environmental and biological cycles that recur with the day/night cycle. It programmes us to sleep at night. In 2019, the International Agency for Research on Cancer (IARC, the specialised cancer agency of the World Health Organization) estimated that about one in five workers worldwide were involved in night-shift work. If the circadian master clock rapidly synchronised to our work patterns as its dominant time cue from the environment, this would not be a problem. Unfortunately, the evidence confirms that it rarely adapts fully to work patterns that are out of step with the day/night cycle.

As a result, shift work often requires people to work through much of the circadian master clock rhythm when they should be sleeping. Most shift workers also choose to switch back to sleeping at night on their days off. Consequently, the circadian timekeeping system not only receives conflicting patterns of light exposure, meal timing, exercise, and social cues on workdays, it also receives unstable patterns of time cues between workdays and days off. When people are required to work at different times across successive days (rotating rosters), the patterns of time cues become even more complex.

A challenge with trying to evaluate the overall impact of shift work on safety, health, and well-being is that there is no standard way of describing work patterns (rosters). Variables include shift start and end times, durations, and sequences (how many shifts in a row and if they rotate), and the length and frequency of breaks within and between shifts. Multiple aspects of work patterns can interact in their effects on the circadian timekeeping system and sleep. Night work clearly represents the worst-case scenario for sleep loss because it requires staying awake throughout the optimum part of the circadian master clock rhythm for sleep. However, late-finishing shifts that delay bedtime and early starts that truncate sleep in the morning can also lead to cumulative sleep debt across a roster.

Given the importance of sleep for safety, health, and well-being, a useful chronobiological definition of shift work is:

> any work pattern that requires you to be awake when you would normally be asleep if you were free to choose your own schedule.

This definition focuses on the physiological impact of shift work – disruption to the circadian timekeeping system and sleep. From a practical point of view, reducing this impact should help reduce the adverse impacts of shift work.

On the other hand, this physiological disruption is clearly not the only thing that can have an adverse impact on shift workers. Working out of step with predominantly day-active family, friends, and society in general can also create significant challenges. For example, in 2016–17 we conducted a national survey of nurses working in six practice areas in New Zealand public hospitals. On a standard question, work patterns caused problems with social life for 49 per cent of nurses, with home life for 38 per cent, with personal relationships for 35 per cent, and with other commitments for 39 per cent.[1]

The challenges of night work

Monitoring the circadian rhythms of shift workers across full roster cycles and days off is logistically complicated and can be burdensome for all concerned (although new genetic biomarker methodologies look promising). As a result, there are only a few studies that illustrate the disruption to sleep and circadian rhythms caused by shift work. The following example (Figure 5.1) comes from data that I analysed for the early NASA study of night-cargo pilots, which used the monitoring system described in Chapter Two. This study illustrates the strong influence of the circadian master clock on sleep timing in a real-world context.[2]

Figure 5.1 summarises core temperature data from 10 night-cargo pilots who had an eight-day duty cycle with short flights up and down the East Coast of the United States crossing no more than one time zone. They worked three consecutive nights (dark grey boxes indicate when they were flying), after which they were flown home, as indicated by the two white boxes on trip day 3. These repositioning

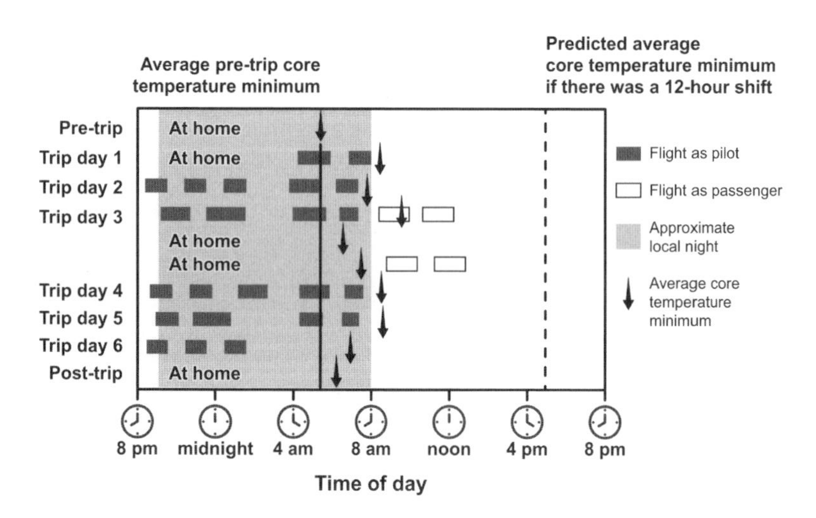

Figure 5.1 Flight patterns and average time of the daily temperature minimum for 10 night-cargo pilots.

('deadhead') flights counted as duty time, but the pilots were flown as passengers. They then had a full night off at home, before deadheading to a hub in a different city to begin another three nights of flying. The downward arrows show the average time of their core temperature minimum each day, as a marker of the circadian master clock rhythm.

When pilots were sleeping at home at night prior to starting this duty cycle, the average time of the minimum in core temperature was 5.20 am. If the circadian master clock had adapted fully when they switched to working at night, this would have shifted 12 hours to 5.20 pm (the dashed vertical line in the figure). The temperature minimum moved slightly later after night duty (8.05 am on average) and did not return completely to its pre-trip timing on the first full night post-trip.

The failure of the circadian master clock to adapt to this work pattern has two important consequences: 1) pilots were working at times in the circadian master clock rhythm when they should be asleep; and 2) they were trying to sleep at times in the rhythm when they should be awake.

Considering the possible safety consequences of working at suboptimal times in the circadian master clock rhythm, there is a saying in aviation that take-off is optional, but landing is not. Landing is a time of high workload and high safety risk. The figure shows that the last landing of each night (end of the last grey box each night) was occurring close to the time of the circadian temperature minimum. This is when sleepiness is strongest and performance capacity on many tasks is around its daily lowest, and it has become known as the window of circadian low (WOCL). In international aviation regulations designed to reduce the safety risk associated with pilot fatigue, the WOCL is defined as from 2.00 am to 6.00 am in the time zone to which a pilot's circadian master clock is adapted.

For the night-cargo pilots in the figure, a high-workload, high-safety-risk phase of flight coincided with a time of reduced pilot performance capacity. There were flight deck observers on all the flights monitored in the NASA studies and no pilot performance issues were observed. However, all the flights were routine – no problems arose that might have challenged the reduced capacity of the pilots to respond. The failure of the circadian master clock to adapt to the work pattern resulted in a reduced safety margin, particularly for the last landing each night.

The challenges of trying to sleep at suboptimal times in the circadian master clock rhythm are summarised in Figure 5.2 (which includes additional data from pilots on a slightly different eight-day trip schedule). The upper panel shows pilots' sleep at home at night when they were synchronised to the day/night cycle prior to the trip.

On average, pilots reported falling asleep at 0.33 am and the core temperature minimum occurred at 5.20 am. Laboratory studies predict that the evening wake maintenance zone, when it is hard to fall asleep (see Chapter Four), is about six to eight hours before the temperature minimum, i.e., around 9.20–11.20 pm. This probably did not affect the pilots' ability to fall asleep at home, which occurred on average after midnight. On average, pilots woke at 8.13 am. Laboratory studies predict that the circadian wake drive, which stimulates wake-promoting brain centres, is most likely to overwhelm the homeostatic pressure for sleep about six hours after the temperature minimum, i.e., around 11.20 am. This would not have woken these pilots up at home because they had already been awake for about three hours.

The lower panel shows pilots' sleep in the morning after night duty (they slept in the morning after 95 per cent of night duty periods). On average, pilots reported falling asleep at 8.55 am and the core temperature minimum occurred at 8.05 am. The predicted time of the evening wake maintenance zone was therefore around 0.05–2.05 am and would not have affected their ability to fall asleep in the morning. However, it appears to have affected their ability to nap in the evening before night duty. Breaks in which they napped

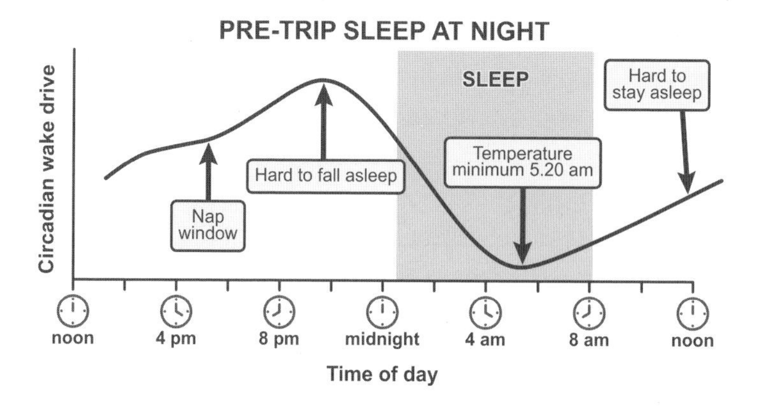

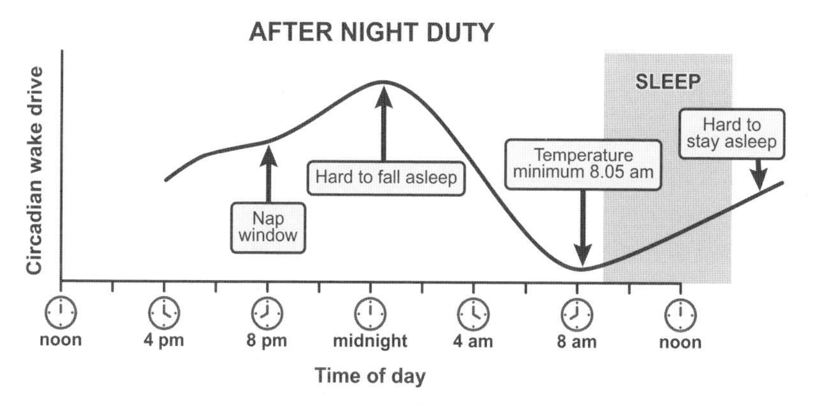

Figure 5.2 Circadian regulation of pilots' sleep at night at home before the trip (upper panel) and in the morning after night duty (lower panel).

before night duty ended between four and seven hours later (around 3.30 am) than breaks in which they only slept in the morning.

On average, when pilots slept in the morning, they woke at 2.13 pm. The predicted peak in the circadian wake drive was around 2.05 pm. Anecdotal reports from crew members indicated that they often woke spontaneously around this time but did not feel well rested. Pilots reported that their daytime sleep episodes were shorter, lighter, less restorative, and poorer overall than pre-trip sleep at night at home. They were also three times more likely to nap on days between night duty but still averaged 1.2 hours less total sleep per 24 hours on duty days. Thus, they were accumulating a sleep debt across successive night duties. This means that their performance capacity would have been lower on each successive night duty. The night off in the middle of the duty sequence provided an important recovery opportunity. Pilots averaged 41 minutes more sleep per 24 hours than pre-trip and 115 minutes more than during daytime breaks between night duty periods.

The times that pilots were eating and exercising when they were working nights may also have generated desynchrony between the circadian master clock and the circadian clocks in their peripheral organs. This could potentially contribute to digestive problems in the short term and increase the risk of becoming overweight or obese in the long term. The long-term health effects of shift work are considered in more detail later in this chapter.

The finding that the circadian master clock did not synchronise to permanent night work is not unique to this study. A review of six studies monitoring the circadian melatonin rhythms of permanent night workers (41 men and 35 women) found that only 3 per cent were fully adapted to their work patterns.[3] (Studies in isolated workplaces that inhibit family and social contact, such as offshore oil rigs and remote fly-in–fly-out mining operations, were excluded.)

Jet lag

When sea or land travel was the only way to move across time zones, the change in environmental time occurred slowly and the circadian master clock could adjust day by day. Jet lag is the consequence of moving to a new time zone at a rate that is too fast for the master clock to adapt during the trip.

Jet lag symptoms vary but include sleepiness at undesired times, being unable to sleep at desired times, feeling hungry at inappropriate times, gastrointestinal complaints (e.g., indigestion, diarrhoea, increased/decreased frequency of bowel movements), mood changes, headaches, and reduced performance capacity. The symptoms are caused not only by the circadian master clock being out of step with local day/night cycle (external desynchrony), but also by internal desynchrony among different organs and circadian rhythms in the body.

For the occasional traveller, jet lag symptoms are a transient inconvenience. As was shown with my trip from Wellington to Montréal in Chapter Two, the circadian master clock can adapt to the abrupt changes in the natural day/night cycle after flights across time zones. The caveat is that you have to stay long enough in the destination time zone to become fully adapted to local time.

Adaptation is generally faster after westward flights than after eastward flights crossing the same number of time zones. One likely reason is that the innate length of the circadian master clock rhythm is longer than 24 hours for most people (see Chapter Two). Adapting to a new time zone after a westward flight requires it to delay (undergo a few cycles longer than 24 hours). In contrast, adapting to a new time zone after an eastward flight requires the circadian master clock to advance (undergo a few cycles shorter than 24 hours).

In general, adaptation to a new time zone is slower when more time zones are crossed. However, sometimes after an eastward time zone shift, the circadian master clock will adapt by effectively going west. For example, instead of advancing six hours to adapt to crossing six time zones east, it will delay 18 hours, effectively adapting by crossing 18 time zones west. This 'resynchronisation by partition', first described by Karl Klein and colleagues in the 1970s,[4] results in much slower adaptation to the new time zone.

Adaptation is also faster with greater exposure to the local time cues (particularly exposure to sunlight, as well as sleeping and eating on local time). I have noticed that I adapt more quickly when I arrive in a new time zone in summer, when there are more hours of sunlight available, than if I arrive at the same destination in winter. Jet lag symptoms subside when all our circadian rhythms arrive in the new time zone and resume their usual relationship to each other and the day/night cycle.

Shift work and jet lag combined

In contrast to occasional travellers, for international airline pilots jet lag is a routine work hazard. Most trip schedules do not allow enough time in each destination for pilots to become adapted to local time. Figure 5.3 is again an example from the early NASA studies where the circadian master clock rhythm was tracked by monitoring the circadian rhythm in core body temperature. The figure shows a trip pattern with six back-to-back transpacific flights separated by about 24 hours in each destination.[5]

The figure is plotted on the pilots' home base time in San Francisco, to which they were adapted before the trip. The light grey bars indicate when pilots experienced local night in the different time zones across the trip. In this trip sequence, there were no consistent 24-hour time cues from the environment.

The figure confirms that the circadian master clock could not synchronise to the rapid sequence of time zone changes and so drifted progressively later (the downward arrows move progressively to the right). Sometimes the average temperature minimum occurred during flights, such as on the return flight from Narita to Singapore on day five, when it occurred just before landing, resulting in a potential reduction in the safety margin. By the time pilots finally returned home to San Francisco, their circadian master clocks had delayed by about 4.5 hours on average, so they had jet lag and had to readapt to their home time zone.

This study was done in the 1980s, and since then aircraft capable of much longer flights have been developed. Very long flights are flown by four-pilot crews, with each pair of pilots taking turns

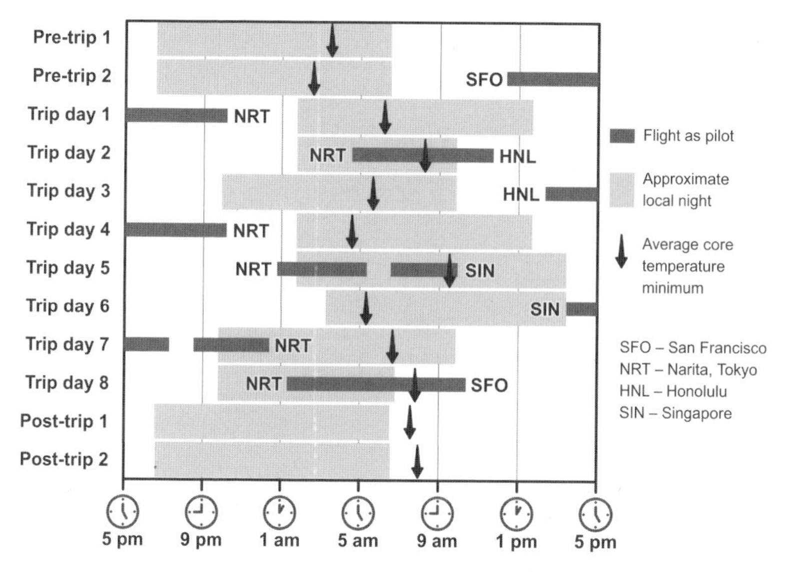

Figure 5.3 Flight patterns and average time of the daily temperature minimum for six long-range pilots (two-pilot crews). San Francisco to Narita (westward), –8 hours; Narita to Honolulu (eastward), +5 hours; Honolulu to Narita (westward), –5 hours; Narita to Singapore (westward), –1 hour; Singapore to Narita (eastward), +1 hour; Narita to San Francisco (eastward), +8 hours.

flying the aircraft or taking a break in a crew rest facility with lie-flat bunks. In 2012, the Sleep/Wake Research Centre monitored pilots in four-pilot crews on a similar trip pattern to the one above, except that the trips began and ended in cities on the US East Coast, i.e., three time zones further east with much longer outbound and inbound flights. It was not possible to track the circadian master clock in these studies. However at specific times across each flight, pilots were asked to complete a validated 10-minute performance test on a hand-held device. It tested how fast they could respond to signals appearing at random intervals on the screen (the psychomotor vigilance task, or PVT, which has been shown to have a strong circadian rhythm).[6] The pilots' responses just prior to the critical descent and landing phase of flight suggested that circadian drift also occurred across these trips. Their total in-flight sleep across the trips also changed in a pattern consistent with circadian drift.

What are the options for reducing the likelihood of circadian disruption for pilots on long-range trips? One school of thought is that it is better to fly to one destination, stay for a day or two, and then fly home. The idea is to try to stay on home base time and so have minimal readaptation after the trip. This way, pilots recover more quickly and can be ready sooner to fly another similar trip. If it works, this should also reduce the amount of circadian disruption they experience in the long term.

In 2019, the IARC extended the definition of night-shift work to include 'transmeridian air travel during the regular sleeping hours of the general population'.[7] For pilots and cabin crew who work on international flights, this needs to be modified to 'the regular sleeping hours in the time zone to which a crew member's circadian master clock is adapted'. Estimating which time zone that is remains a significant scientific and operational challenge in long-range airline operations.

Shift work and safety

One of the challenges in evaluating the effects of shift work on safety is that many studies do not describe critical aspects of work patterns, such as the timing and sequences of shifts and recovery sleep opportunities. In addition, very few studies track information on people's work patterns over time.

A further challenge is that a person's functional capacity is not the only factor that affects safety. It also depends on what they are being asked to do, as well as the other hazards and safety defences present. Recall that the two-pilot night-cargo crews in the NASA study were able to safely complete routine flights when they were experiencing the combined effects of cumulative sleep debt and working through the least functional time in the circadian master clock rhythm. The presence of the second pilot is a key safety defence, as is air traffic control. Nevertheless, if a crew had needed to respond to any complex critical events, their reduced functional capacity could have become a deciding factor in the outcome.

Another illustration of the importance of context is to think about the safety risks associated with a person who is so sleepy that they are about to fall into a brief, uncontrollable microsleep. If that person is alone in a truck cab at night and driving at 100 kilometres per hour, during a 10-second microsleep they will travel 278 metres and there is a relatively high risk of a serious motor vehicle crash. The likelihood of multiple injuries or deaths may be lower if he/she is driving on a rural highway, where there is less oncoming traffic, than if he/she is in morning rush-hour traffic after a long intercity drive. By comparison, if an airline pilot in a four-pilot crew on a very long flight falls into a microsleep on the flight deck, the likelihood of a plane crash is extremely low. There is another pilot in the other seat and if they are in the cruise portion of the flight, the aircraft is probably being flown by the autopilot. There are two other pilots taking their turn in the crew rest facility, and the captain in charge of the flight can allocate rest breaks according to the needs of each pilot on the day. There are international and (in most countries) national requirements for airlines to manage pilot fatigue and for pilots to receive fatigue management training. In other words, there are multiple layers of safety defences in place to reduce the likelihood of a pilot being extremely sleepy on the flight deck, as well as strategies to manage the situation if it does arise.

Population-based surveys on shift work and safety typically rely on asking people about broad categories of work patterns. For example, in a survey of over 15,000 New Zealand blood donors in paid employment, participants were asked to identify their work pattern in one of five categories: 1) daytime with no shifts; 2) irregular hours; 3) rotating shifts without night work; 4) rotating shifts with nights; or 5) permanent nights.[8] Shift work – defined as rotating shifts with or without nights, or permanent nights – was worked by 3119 people (21 per cent) with 2331 (15 per cent) working at least one night per week. Rotating shift workers with or without nights were almost twice as likely as day workers to report a workplace injury in the last 12 months that had required treatment from a doctor. The analyses accounted for the influence of sex, education, hours worked, smoking, daytime sleepiness, sleep difficulties, headaches, body mass index, and occupational group. (Some occupational groups have higher risk of workplace injury because of the work environment and the nature of their tasks. New Zealand examples include deck hands on fishing vessels and logging crews in forestry.)

A French cohort study followed over 3000 employed and retired workers from a range of occupations across 10 years, starting in 1996.[9] They were asked to identify their work pattern (currently, in the past, or never) in four categories: their work 1) involved rotating shifts (e.g., alternating morning, afternoon, and night shifts); or 2) did not allow them to go to bed before midnight; or 3) meant they had to get up before 5 am; or 4) prevented them sleeping during the night (night work). Participants were 32, 42, 52 and 62 years old at the time of the first measurement, when 48 per cent had experience of shift work. Cognitive speed and memory were tested at five-year intervals. Cognitive impairment increased with increasing years of shift work, although the speed at which people responded did not change. Those who had worked shift work for more than 10 years showed a decline in cognitive function equivalent to 6.5 years of age-related decline in the cohort. Cognitive functioning returned to an age-equivalent level after leaving shift work, but this took at least five years. The authors highlight the potential reduction in shift workers' safety and quality of life, as well as the increased safety risk for society. They advocate medical surveillance of shift workers, especially those who have remained in shift work for 10 years or more.

Nursing is a profession where there has been a large amount of research on shift work and safety. With regard to patient safety, incidents are more likely on the night shift, relative to the number of patients/procedures. Clinical errors are reported more frequently by nurses working rotating shifts compared with day shifts or evening shifts, and by those working night shifts compared with day shifts. In studies where nurses kept daily logbooks for a month, restricted sleep on workdays was associated with greater likelihood of reporting errors. In our 2016–17 national survey of nurses working at least 30 hours per week in New Zealand public hospitals, 31 per cent could recall making a fatigue-related clinical error in the last six months.[10] The likelihood of recalling an error was related to their work patterns in the last two weeks. Errors were more likely with more total hours worked, more night shifts, more shift extensions of at least 30 minutes, and unplanned roster changes.

With regard to the safety of nurses themselves, there is evidence from other studies that long work hours and shift work increase the risk of needlestick and other work injuries; musculoskeletal disorders in the neck, shoulders, and back; as well as nurses reporting motor vehicle accidents and near-misses returning home from a shift. In our 2016–17 national survey, 32 per cent of nurses reported that, since becoming a nurse they had fallen asleep driving home from work and 65 per cent had felt close to falling asleep at the wheel in the last 12 months. Drowsy driving associated with nurses' work patterns represents a safety risk to other road users as well as to the nurses themselves.

Since the early twentieth century, the transport sector (rail, road, aviation, and maritime) has grappled with a recognised safety issue understood intuitively as 'operator fatigue'. The traditional approach to managing operator fatigue was to impose regulatory limits on maximum duty times and minimum breaks. However, as the effects of circadian desychrony and sleep loss became known, and vehicles and operations became increasingly complex, it was recognised that prescriptive duty/rest limits alone are often insufficient to manage the safety risks associated with operator fatigue.

In 2009–10, the UN agency responsible for developing standards for the global aviation industry (the International Civil Aviation Organization, ICAO) assembled a taskforce with international representation from regulators, airlines, unions, and scientists, of whom I was one. Our task was to update the global regulatory framework for managing pilot and cabin crew fatigue in commercial aviation, but first we had to come up with a scientifically valid and operationally useful definition of fatigue. This is my translation into slightly less technical language:

> Fatigue is a physiological state of reduced physical and mental performance capability caused by sleep loss, extended wakefulness, circadian phase, and/or workload (mental and/or physical) that can impair a person's alertness and ability to perform efficiently and safely in the workplace.

This definition identifies the human problem (reduced performance capability), its four most common causes, and its consequences for workplace productivity and safety. Defining fatigue as a physiological state also clarifies that a fatigue-impaired person is unable to function at her/his best, rather than unwilling or unmotivated. Once you agree on what fatigue is, what causes it, and how it can impact on safety, you can develop systematic strategies for managing it.

The US National Transportation Board (NTSB) is an independent government agency that investigates accidents in all modes of transportation. Since 1989 it has made recommendations to the US Department of Transportation aimed at reducing fatigue-related accidents in all modes of transportation. This was still listed as a priority 30 years later in the NTSB's 2019–20 list of the 10 most wanted safety improvements in transportation. In support of this finding, the Board noted that: since 1982, 300+ aviation accidents in the US have been associated with fatigue; in 2016–17, it investigated seven maritime accidents in which fatigue was a factor; an estimated 6400 fatal crashes on US roads each year involve a drowsy driver; and in 2001–12 there were seven major rail investigations in which the Board identified fatigue as a probable cause, or a contributing factor, or as a finding.

Applying science to improve safety

Scientific understanding about the circadian master clock's failure to synchronise to work patterns, and the consequences for sleep and waking performance capability, provides some principles for improving shift work. These are summarised in the following table.

Table 5.1 Scientific Principles for Roster Design

Scientific principles	Shift work principles
Circadian challenges of shift work	
The circadian master clock rarely adapts fully to work patterns that do not allow normal sleep at night.	People working during their usual sleep time will not be able to function at their best. People trying to sleep outside their usual sleep time will have more difficulty getting enough sleep.
Importance of sleep	
Getting enough sleep (both quantity and quality) on a regular basis is essential for restoring the brain and body.	Sleep opportunities matter, not just rest breaks. A 10-hour break from 9 pm to 7 am is a much better sleep opportunity than a 10-hour break from 9 am to 7 pm.
Dynamics of sleep loss and recovery	
The effects of sleep loss build up across multiple days (sleep debt), eventually resulting in unintended microsleeps. As sleep debt builds up, how sleepy we feel is not a reliable indicator of how we are functioning. At least two consecutive nights of unrestricted sleep are needed for recovery from sleep debt.	For recovery from sleep debt, rosters need to include regular breaks of at least two nights off in a row (recovery breaks). This is not the same as 48 hours off. For most people 48 hours off starting at midnight allows only one night of unrestricted sleep. How often recovery breaks are needed depends on how fast sleep debt is building up. When shifts overlap more of a person's usual sleep time at night, then sleep debt builds up faster.

Continued overleaf

Table 5.1 *(continued)*

Fatigue and safety risk	
The safety risk associated with a fatigue-impaired person in the workplace depends on what she/he is being asked to do, the other hazards present, and the other safety defences present.	Limits on shift lengths, number of consecutive shifts, breaks, etc., aim to limit the level of fatigue of people at work. They do not address the differences in risk associated with fatigue-impaired people doing different jobs at different times.

To reduce the safety impact of shift work, scientific understanding alone is insufficient.[11] It needs to be integrated with workplace expertise. People with workplace experience have vital knowledge about the specific safety risk(s) associated with a fatigued individual in their context, as well as about workable strategies for reducing risk. Responsibility for designing and managing shift-work systems in larger organisations often involves coordinating workplace experience with health and safety expertise and management knowledge about organisational requirements, expectations, and constraints.

Managing shift work must be a shared responsibility – what people choose to do outside of work can improve or exacerbate the effects of work demands. Basic education about how shift work can disrupt both sleep and waking function, and strategies for reducing these effects, is vital for all parties.

New approaches for managing shift work are being developed that incorporate scientific and safety management principles with workplace expertise and legal requirements, for example for airline pilots[12] and hospital-based nurses.[13] They focus primarily on the short-term goal of improving safety. If we can reduce circadian desynchrony and sleep loss in the short term, this should help reduce its long-term adverse effects on health. This will not, however, be a magic cure-all for getting rid of the consequences of trying to override our circadian timekeeping system.

Shift work and health

Three main types of evidence are converging to create a concerning picture of the possible impacts of shift work on health: 1) large, population-based studies tracking the health of shift workers and non-shift workers across time; 2) human time isolation laboratory studies that cause short-term circadian disruption to establish whether it has effects that could contribute to adverse health outcomes; and 3) experiments with animals that explore the genetic, metabolic, and health effects of long-term simulated shift work. There is still scientific debate in this area, but looking carefully at the increasing quantity and quality of the evidence, I am convinced of the need for more effective action.

One of the most compelling population-based studies is the US Nurses' Health Study, which began in 1976 and to date has included over 280,000 nurses.[14] It has compared over time the health of nurses who work at least three nights per month with nurses who do not work nights. After five years, nurses working nights have significantly higher mortality rates from all causes and from cardiovascular disease. After 15 years they are more likely to have died from lung cancer, to have had a stroke, and to have developed colorectal cancer. They are also more likely to develop breast cancer and type 2 diabetes. The Danish Nurse Cohort Study has also followed the health of nurses working night and evening shifts compared with those working day shifts only.[15] Nurses working nights have an increased risk of developing diabetes and are at increased risk of dying from cardiovascular disease, type 2 diabetes, Alzheimer's disease and other types of dementia.

People working night shift and rotating shift patterns are more likely to have metabolic syndrome – a cluster of risk factors for cardiovascular disease and type 2 diabetes that includes obesity, high blood pressure, and high levels of triglycerides, cholesterol, and glucose. Both men and women working rotating shifts develop metabolic syndrome faster than dayworkers, after controlling for other risk factors. Multiple cohort studies have found increased risk of type 2 diabetes in shift workers compared to day workers in the same industries.

How could shift work increase disease risk? As you might expect, this is a very active area of research, but it is complex because diseases have multiple causes and individuals have different combinations of risk factors that evolve over time. There are also multiple possible mechanisms for the adverse effects of shift work on health including the cumulative effects of repeated sleep loss (as discussed in Chapter Four), repeated circadian desynchrony, and unhealthy diet. A major challenge is that shift workers, particularly night workers, can experience the combined effects of all three of these factors.

For the adverse effects of shift work on metabolic health, laboratory evidence suggests at least three possible mechanisms: sleep loss changing the balance of key hormones regulating appetite so that people eat more; eating at times in the circadian master clock rhythm when the digestive processes are suboptimal; and decreased overall energy expenditure when people are sleeping and eating at the wrong times in the circadian master clock cycle.[16] Laboratory studies also show that sleep loss causes disrupted glucose metabolism and reduced insulin sensitivity, which can be reversed with sufficient recovery sleep in otherwise healthy people.

An elegant new series of time isolation laboratory studies from the Harvard Medical School team has started teasing apart the effects of sleep loss and circadian disruption.[17] Zitting and colleagues compared the effects of living for three weeks on a 28-hour light/dark schedule, which caused repeated disruption to the circadian timekeeping system, with living for three weeks synchronised to a regular 24-hour light/dark cycle. Both lighting regimens allowed enough darkness to minimise sleep loss. Half of the participants were given a high-fat diet while the other half received a low-fat diet. Across the 28-hour light/dark cycle regimen, people on the high-fat diet increased their body fat content and developed impaired glucose tolerance and insulin sensitivity, but these effects did not occur in the people on the low-fat diet. Across the 24-hour light/dark cycle, neither group showed these changes. In each of the light/dark cycles, sleep was comparable in both groups. Similar results were found in parallel experiments with young and old mice. The findings suggest that circadian disruption alone (i.e., without sleep loss) can have adverse effects on metabolism and that reducing dietary fat may help protect against these effects.

Another possible mechanism for the adverse health effects of shift work that is receiving a lot of attention is the idea that exposure to light at night might play a role. It could possibly be acting in two ways: by resetting the circadian master clock and increasing desynchrony in the circadian timekeeping system; and/or by suppressing the synthesis of melatonin. Several studies have found evidence of lower melatonin levels among both male and female night workers (by measuring a melatonin metabolite in urine). The circadian rhythm of melatonin secretion across 24 hours is flattened (lower amplitude) in shift workers versus non-shift workers. Decreased morning melatonin levels in humans are associated with an increased risk of hypertension and type 2 diabetes. Administering melatonin can also suppress tumour growth in mice and in human breast cancer cells,[18] and melatonin may play a role in maintaining a healthy cardiovascular system.

The scientific evidence has in fact been building for decades, as has concern particularly about night work. In 1990, the International Labour Organization (ILO, a UN agency that develops policies and sets standards promoting decent work conditions), passed a Night Work Convention (C171), with night work defined as working at least seven consecutive hours including midnight to 5 am. Provisions include that night workers can request a health assessment without charge and receive advice on how to reduce or avoid health problems associated with their work. If they are unfit for night work, they should be transferred whenever practicable to a similar job for which they are fit. An alternative to night work should be provided for women for at least eight weeks before birth, for a minimum of 16 weeks total before and after birth, and longer if deemed medically necessary for the health of the mother or child. In no case should a worker be disadvantaged financially or otherwise by these provisions. Thirty years later, only 17 out of 187 UN member states have ratified this convention.[19]

In 2007, the IARC classified shift work involving circadian disruption as 'probably carcinogenic to humans' (i.e., a level 2A carcinogen, with level 1 being 'carcinogenic'). This classification was made 'on the basis of sufficient evidence in experimental animals and limited evidence of breast cancer in humans'. In 2009, Denmark began compensating breast cancer as a work-related illness for women who developed it after long-term night work and were otherwise at low risk – for example, they had low alcohol consumption and no family history of breast cancer. Compensation is paid through employers' insurance schemes.

In 2019, the IARC convened a new working group of 27 scientists from 16 countries to revisit the 2007 conclusions based on new research.[20] This group revised the classification to say that night-shift work (rather than shift work in general) is probably carcinogenic to humans (a level 2A carcinogen). The evidence on which this was based was also revised, namely limited evidence of cancer in humans, sufficient evidence of cancer in experimental animals, and strong mechanistic evidence in experimental animals.

I remember sitting entranced in a lecture theatre at Harvard Medical School in the early 1980s watching one of the first Space Shuttle launches. During my first period of working at NASA (1983–85), I had the memorable experience of watching the *Challenger* land at Edwards Airforce Base in the Mohave Desert. It appeared suddenly, travelling incredibly fast, accompanied by sonic booms that startled hundreds of birds into the air. When it eventually landed, I was astonished by how high its nose was to help reduce speed as it headed in.

On 28 January 1986, watching television in Toulouse, I saw the horrific footage of the *Challenger* breaking apart 73 seconds into its tenth flight, killing all seven crewmembers. My longtime colleague and friend Curt Graeber was part of the team that investigated how human factors contributed to this catastrophe. The immediate cause was the burning through of the O-rings in the Shuttle's right solid rocket booster, leading to the explosion of the external fuel tank. However, the Presidential Commission on the Space Shuttle Challenger Accident also cited the contribution of human error and poor judgement related to sleep loss and shift work during the early morning hours.[21]

The decision to launch had been taken by key managers at Marshall Space Center the evening before, against the advice of Morton Thiokol engineers, who were concerned about the potential for the O-rings to fail in the unusually cold temperatures prevailing at the launch site. Certain key NASA managers had obtained less than two hours' sleep the night before and had been on duty since 1 am that morning. The Presidential Report noted that 'time pressure, particularly that caused by launch scrubs and turnarounds, increased the potential for sleep loss and judgment errors' and that working 'excessive hours, while admirable, raises serious questions when it jeopardizes job performance, particularly when critical management decisions are at stake'.[22]

Clearly, many aspects of twenty-first-century life run counter to our complex and carefully tuned circadian timekeeping system. There is a broad range of reasons why organisations operate 24/7. At one end of the spectrum, there are the essential services on which society depends, such as healthcare, fire and emergency, and the police. Theoretically, at the other end of the spectrum there are organisations whose choice to run 24/7 is primarily motivated by increasing the profit to shareholders.

There is a pervasive lack of awareness of the price we are likely to be paying at the personal, family, community, national, and international levels. We need more sophisticated science to understand the mechanisms and develop effective strategies to reduce the consequences for our health, safety, and well-being. We also need a much wider and well-informed debate about the choices we are making by creating businesses and lifestyles that conflict with our circadian timekeeping system.

Chapter Six

Our Home Planet

Modern lifestyles are clearly having adverse effects on our safety, health, and well-being because they generate conflict with our innate circadian timekeeping system, compounded by our failure to value sleep. Returning full circle to Chapter One, all cell-based life forms on Earth have circadian rhythms. Many also have innate biological clocks that are adaptations to the environmental cycles generated by the orbit of the moon around Earth and the orbit of Earth around the sun. This final chapter looks at some of the adverse effects that modern human lifestyles are having on other organisms and ecosystems. It also considers some unanticipated challenges we may face if we want to take our Earth-based biology elsewhere in the solar system.

What we are doing to Earth's light environment

Until very recently in the history of life on Earth, the day/night cycle, the phases of the moon, and seasonal changes in photoperiod have been reliable time cues. We have changed that markedly in the last two centuries.

Based on satellite imagery, in 2014 it was estimated that artificial light at night affected 11.4 per cent of terrestrial and 0.2 per cent of marine areas of the globe, although with large geographical variation. This environmental change is more extensive and happening faster than any other change caused by humans, including the changes in temperature, carbon dioxide and habitat.[1]

The spectrum of artificial light at night is different from moonlight or starlight. Although it varies with the technologies used and in different areas, there has been a general trend towards the use of 'whiter' light sources, often with a strong component in the blue portion of the spectrum, especially in commonly used light-emitting diodes (LEDs). A wide variety of plants and animals are sensitive to blue light, which has a role in regulating the circadian timekeeping system in many species.[2] Blue light scatters in the atmosphere more than other wavelengths, so at night it produces artificial sky glow, which can extend hundreds of kilometres from the source and is not blocked by obstacles on the ground. Sky glow reduces the visibility of the universe for Earth-based optical astronomy. Clouds also reflect artificial light at night back to Earth.

Many animal species are nocturnal (about 39 per cent of vertebrates and more than 60 per cent of invertebrates), and they tend to be less well studied than diurnal animals.[3] Some species are attracted to light sources at night, including many insects and migrating diurnal birds at sea, which are attracted to bright light from lighthouses, ships, and oil rigs. This can lead them to exhaustion and death. Sometimes the attraction is indirect, such as bat species feeding on insects around street lights. On the other hand, some bats and bird species avoid artificial light at night. This behaviour is harder to detect and so may be more common than is currently thought.[4]

As well as these immediate effects, artificial light at night can have longer-term effects on the physiology of animals and plants, including resetting their circadian and circannual timekeeping systems. Artificial light at night advances the dawn song of some birds and can alter the timing of seasonal breeding, possibly through suppressing melatonin synthesis. Many plants have phytochrome photoreceptors that are sensitive to the amount of red/infrared light. These photoreceptors measure day length and time key events such as budburst, flowering, and bud set to coincide with favourable environmental conditions. One study has matched the times of budburst on four deciduous tree species across the United Kingdom with both satellite imagery of artificial light at night and average spring temperatures across 13 years. Budburst occurred up to 7.5 days earlier in brighter areas and this could not be explained by local or annual differences in temperature.[5] Another commonly observed phenomenon is that trees close to street lights often shed their leaves later. This is an indicator of slower physiological preparation for winter dormancy, which can put them at greater risk for frost damage.[6] Young trees naturally grow more vigorously and for longer than older trees, and so are more prone to cold injury as a result of growth prolonged by artificial light at night.

Much of the research in this area has focused on specific animals and plants in urban environments where artificial light at night is not the only human activity that affects ecosystems. A fascinating Dutch research project has set up different artificial lighting environments with white, green, or red light along the edges of forests, to monitor the longer-term effects on a less-disturbed natural ecosystem.[7] The lit areas are being compared with an area with no artificial light at night. All the sites are within nature reserves and the lights come on at sunset and go off at sunrise.

Observations from the first three years include that wood mice are much less active (not caught on cameras) in all of the areas with artificial light compared to the dark area. They are known to be less active in moonlight, probably to avoid predation.

Pipistrelle bats are foraging (as measured by recording their calls) much more in the green and white light areas, probably because of the insects attracted to these lights. Bat species that are less agile fliers (less able to avoid predators) avoid light at night and so could be having their access to food reduced.

Based on six years of data, female great tits (*Parus major*) have been found to lay their eggs significantly earlier in nests in the white and green light areas (with a similar trend in the red light area) versus the dark area.[8] However, this effect was found only in late cold springs, not in warm early springs. It may be that rising temperatures have a greater effect than light pollution on the timing of breeding in these birds, possibly through effects on other species. Much more data is needed to confirm this.

Interactions between the effects of artificial light at night and climate change are an area of concern. As the rate of climate change increases, it has been suggested that many plants, especially trees and shrubs with long generation times, may have insufficient time to adapt to the combined effects of shifts in their biological time-keeping systems due to artificial light at night as well as changes in temperature.[9]

Marine ecosystems have received less attention than land-based ecosystems when it comes to the effects of artificial light at night.[10] Coastal areas have light pollution from towns and cities. Further out to sea, as noted above, artificial light comes from shipping, oil platforms, and light fisheries. I can recall seeing what looked like floating towns of squid boats off the New Zealand coast at night.

Disorientation by artificial light at night prevents or delays some marine turtle hatchlings reaching the sea, leading to reduced nesting in artificially lit areas.[11] Some fish species are attracted into lit areas of estuaries, and nocturnal bird strikes on vessels at sea were noted earlier. In terms of biological timekeeping, artificial sky glow masks the cycle of moonlight intensity, which is an important time cue for spawning in many marine species. Animal plankton migrate to the surface at night to feed on plant plankton and to avoid diurnal predators. This is arguably the largest daily migration of biomass on the planet. The faecal pellets excreted when animal plankton go back down into the ocean during the day are a major pathway in the carbon cycle, which is likely to be disrupted by artificial light at night. A great deal more research is needed to determine the scale of the problems in marine ecosystems and to develop strategies to reduce them.

Solutions are available now that can reduce the extent and impact of artificial light at night on the biosphere. What is needed is the will and resources to implement them, together with open, informed debate about the costs and benefits of our nocturnal activities that are contributing to this major environmental change. One of my most magical experiences ever was a recent visit to the dark sky sanctuary of Aotea Great Barrier Island off the north-east coast of New Zealand. I walked on the beach by starlight, saw the Magellanic Clouds beyond our galaxy with my own old eyes, and was overwhelmed by the number of stars in the Milky Way.

The challenges in trying to understand the effects of artificial light at night highlight for me the complexity of natural ecosystems and how much we still have to learn using our current scientific paradigms. We also seem to have lost vital old knowledge about the natural world. For a time, I lived in a rural area of France that was originally inhabited by Neanderthals. I was at home with young children, and an old woman from the village, who had been the nanny of my husband and his sisters, introduced me to two of the local woodland areas. Walking with Juliette, I learned where the mycelium grew for different types of mushrooms, and the time of year and phase of the moon when each would produce delicious food. I also learned where and when to forage for other delicacies like wild asparagus and salad greens. She planted according to the phases of the moon and had an incredibly productive garden. I now wish I had been a better student.

What we are doing to our own light exposures

As described in Chapter Two, the circadian master clock in humans and other mammals synchronises to the day/night cycle primarily by tracking blue light via specialised cells in the retina. These cells also communicate with other brain areas unrelated to vision. Via these pathways, blue light, whether artificial or from natural daylight, can increase daytime alertness, ability to concentrate, and performance capacity.[12]

Our use of artificial blue light has increased rapidly with the move from incandescent to LED light sources for domestic lighting and with the explosion in our use of light-emitting screens, including computers, smartphones, tablets, and light-emitting e-readers. Globally, there is also a transition to using LEDs in street lighting (and as noted earlier, white LEDs have a higher proportion of blue light than older light sources such as high-pressure sodium street lights).[13] We have also greatly reduced our exposure to daylight, particularly in urban environments and with the electronic entertainments that entice us indoors. Surveys in industrialised countries suggest that people may spend up to 90 per cent of their time indoors.[14]

In addition to being free, daylight has numerous advantages over artificial light. Vision is our most developed sense and daylight supports good vision. It has been proposed that lack of exposure to daylight may be a cause of the current increase in myopia (short-sightedness) among children. Sunlight on the skin is also important for vitamin D production. Unlike our artificial light sources, daylight is very dynamic. It changes in intensity, colour, diffuseness, and direction both within and across days, with the weather, and throughout the year. Seasonal changes in daylight affect brain function, mood, and depression. Apart from their physical properties, sunlight and shadows have had aesthetic and spiritual connotations across human history. When asked, people generally prefer daylight to artificial light.

Historically, exposure to bright blue daylight and avoidance of light at night provided strong, predictable time cues to our circadian timekeeping system. The combination of less exposure to daylight and more exposure to artificial light in the evening delays the circadian master clock cycle, leading to later sleep times and increased difficulty waking up in the morning. One solution that is gaining favour is to design buildings that allow more exposure to daylight. Sophisticated artificial lighting systems are also being developed that mimic the dawn and dusk changes in the spectrum and intensity of light, or even weather patterns, although these have the disadvantage that they require an energy supply.

Taking our Earth-based biology elsewhere

On 20 July 1989, marking the twentieth anniversary of the *Apollo 11* moon landing, US president George H. W. Bush announced plans for the Space Exploration Initiative, which included constructing a space station, human habitation on the moon, and human exploration of Mars. I was working at NASA and for a brief, heady period I veered into astronautics. I wondered what would happen to our biological timekeeping systems if we removed them for long periods from Earth's geophysical cycles. There was already evidence that synchronisation, particularly with the 24-hour day/night cycle, was essential for maintaining mental and physical health, and for optimal waking function.

Clearly a Mars mission would be the biggest challenge (for many reasons, in addition to our Earth-based biological timekeeping). Mars rotates on its axis and has a 24.6-hour day/night cycle. For most people, the circadian master clock could adapt to a 24.6-hour day/night cycle, but bathing in direct sunlight on the Martian surface is not an option. Mars has no global magnetic field to deflect high-energy particles and its atmosphere is much thinner than

Earth's, so particles from the sun reach the Martian surface, as do high levels of cosmic radiation. Spacesuits would have to be worn to breathe and for radiation protection. Gravity on Mars is also only 38 per cent of Earth's. Mars is tilted on its axis by 25 degrees (compared to Earth's tilt of 23.4 degrees), so Mars has seasons. However, it takes 687 Earth days to orbit the sun, so its seasons are longer. Earth's orbit around the sun is almost circular, so our seasons are about the same length (Earth days in the northern hemisphere: spring 93, summer 93, autumn 90, winter 100). In contrast, the orbit of Mars is elliptical, so the seasons are of different lengths (in Martian days: spring 194, summer 178, autumn 142, winter 154). Earth-based annual biological timekeeping is based on photoperiodism and circannual clocks that typically have innate periods shorter than an Earth year, so it seems unlikely that they could be synchronised by the very long and very different pattern of the Martian seasons.

The length of the trip from Earth to Mars depends on the distance between them, which varies because they go around the sun at different speeds. A human mission to Mars means sending astronauts into interplanetary space for a minimum of a year, even with a very short stay on Mars. During the trip, should astronauts be synchronised to the 24-hour Earth day, or the 24.6-hour Martian day, or switch somewhere on the way? They may well work in a two-shift system. Artificial time cues will have to be provided. Light/dark cycles might work, although, oddly enough, there is some evidence that the phase-shifting effects of light on the circadian master clock may be different in microgravity.[15] Timed exercise might help, and the astronauts would need to exercise to maintain cardiovascular fitness, muscle strength, and bone mass in microgravity. Knowledge about when to do different types of exercise to synchronise the circadian master clock on Earth is incomplete, so there is a lot to learn about how to use timed exercise to minimise circadian desynchrony in microgravity.

Figure 6.1 NASA's 1.8 kg Ingenuity Mars helicopter photographed from the Perseverance rover on 16 April, 2021. Ingenuity has successfully flown multiple short missions in the thin Martian atmosphere. Photo: NASA/Jpl-Caltech/Asu

A series of Mars mission simulation studies was carried out by the State Scientific Center of the Russian Federation Institute for Biomedical Problems (IBMP), including a 520-day full mission simulation.[16] A multinational crew of six healthy men were confined from 3 June 2010 to 4 November 2011 in a pressurised facility with a layout and volume comparable to a spacecraft. This consisted of interconnected modules equipped with life support systems and an artificial atmosphere. An additional module simulated conditions on the surface of Mars. Daily activities simulated those of crews on the International Space Station, including maintenance work, scientific experiments, and exercise. The mission included the simulation of communication delays ranging from eight seconds to 736 seconds as the spacecraft moved away from Earth, and a Mars landing with extra-vehicular activities (EVAs) between days 257 and 265. The crew worked a cycle of five days on, two days off, except for special events such as simulated emergency scenarios. They had control over the lighting and their food intake and exercise. For unknown reasons, they chose to lower the lighting levels progressively across the mission.

Actigraphs worn by the crewmembers showed that on average they became increasingly sedentary across the mission. However, there were marked differences between individuals. One had high waking activity levels and adequate sleep to maintain fast response rates on the psychomotor vigilance task (PVT) throughout the mission. Another had low waking activity levels and short, poor-quality sleep. His PVT performance suggested that he experienced chronic partial sleep deprivation throughout the mission. Four crewmembers maintained a 24-hour sleep/wake cycle with one sleep period at night. Another crewmember maintained a 24-hour sleep/wake cycle but had split sleep, and his daytime sleep increased across the mission. The remaining crewmember had a sleep/wake cycle that lengthened across the mission from 24.7 hours to 25.1 hours. Compared to the others, the latter two crewmembers slept later during the mission than at home pre-mission, as well as having longer sleep during the mission. As result, they were asleep when other crewmembers were awake (or vice versa) for a total of 2498 hours, or 20.1 per cent of the mission. On a real mission, this would certainly affect crew cohesion and productivity and reduce the safety margin.

A 2016 review of spaceflight studies in near-Earth orbit confirms that circadian desynchrony occurs due to schedule constraints and to insufficient or mistimed light exposures.[17] As expected, desynchrony is associated with restricted sleep and increased use of sleep medications. The challenge, as on Earth, is being able to accurately monitor the circadian master clock rhythm to know when to administer countermeasures.

We still have a lot to learn about our Earth-based biological timekeeping systems.

Belonging to Earth

While writing this book, I have often pondered how we became so isolated from the biological timekeeping systems that are integral to how we function and feel. Building dwellings is an ancient adaptation to shelter us from environmental change. With aggregation into larger and more complex villages, towns, and cities came decreasing awareness of how natural ecosystems change with the day/night cycle, the lunar cycles, and the seasons. Somewhere along the way, the illusion took root that we could control everything on the planet with our superior intellectual capabilities and our technological cleverness. With the invention of artificial light, fast aeroplanes, and the internet, we thought we were conquering time. Soon we would be living on other planets and medical breakthroughs would enable us to live forever.

Then (at least for me) along came chronobiology, and with it a realisation that we might not be quite as clever as we think. We can't just divorce ourselves from the geophysical cycles of Earth; they are genetically programmed into us. Life on our planet works in four dimensions, not three. For the future of ourselves, our children, and all life on our planet, we need to renew our acquaintance with biological time and the environmental cycles that govern it.

Glossary

Actigraph A watch-sized device that is worn on the non-dominant wrist and continuously detects movement. The movement patterns are interpreted via a validated computer algorithm to estimate when the wearer was asleep and awake.

Biological clocks Internal timekeeping systems that have evolved in different organisms in response to the predictable changes in Earth's physical environment. These include adaptations to the day/night cycle, the tides and the monthly cycles of the moon, and the annual cycle of the seasons.

Chronobiology The scientific discipline that focuses on understanding how biological clocks work and the adaptive advantages they bring in different environments. Human chronobiology also has a large focus on the consequences of trying to override our innate timekeeping systems.

Chronotypes Differences in when we prefer to sleep and undertake various waking activities, classified across a spectrum from morning-types to evening-types.

Circadian master clock In mammals including humans, the circadian master clock is located in the suprachiasmatic nucleus (SCN) in the hypothalamus area of the brain. It is synchronised by environmental light and sends timing information to biological clocks elsewhere in the brain and in organs throughout the body, and in return receives feedback from some of them.

Circadian rhythm Without any time cues from the day/night cycle, a circadian rhythm repeats roughly every 24 hours (Latin *circa* – about, *dies* – day).

Circadian timekeeping system A complex network of components and control loops that coordinates circadian rhythms generated by cells, tissues, and organs throughout the body, and keeps them in step with each other and the day/night cycle.

Circalunar rhythm Without any time cues from the lunar cycle, a circalunar rhythm repeats roughly every 28 days (Latin *circa* – about).

Circannual rhythm Without any time cues from the seasonal cycle, a circannual rhythm repeats roughly every 12 months (Latin *circa* – about).

Circatidal rhythm Without any time cues from the tide cycle, a circatidal rhythm repeats roughly every 12.4 hours. (Latin *circa* – about).

Cycle In this book, the word 'cycle' refers to a regular fluctuation in the external environment (day/night cycles, tide cycles, lunar cycles, seasonal cycles). This terminology is used to differentiate external 'cycles' from internal 'rhythms' (see the definition of rhythm).

Evening wake maintenance zone The few hours before normal sleep time when the circadian master clock is very strongly promoting wake, which makes it very difficult to fall asleep earlier than usual.

Homeostatic sleep drive A physiological drive for sleep that increases across hours awake and decreases across hours asleep.

Phase Where a biological clock or rhythm is up to in its repeating pattern. Easy phases to identify are the peaks and troughs in each oscillation.

Phase advance A phase advance temporarily shortens the repeating pattern of a rhythm so that it moves earlier. For example, when it is noon in my home in Wellington (New Zealand), it is already 3 pm in Los Angeles. If I fly eastward to Los Angeles, my circadian master clock must speed up temporarily and lose three hours to get onto Los Angeles time. This is a phase advance.

Phase delay A phase delay temporarily lengthens the repeating pattern of a rhythm so that it moves later. For example, when it is noon in my home in Wellington (New Zealand), it is only 7 am in Perth (Western Australia). If I fly westward to Perth, my circadian master clock must slow down temporarily and add five hours to get onto Perth time. This is a phase delay.

Polysomnography The accepted gold standard method for tracking the internal structure of sleep. Surface electrodes on specific sites record electrical activity from the brain, the chin muscles, and eye movements. Changes in these three types of electrical signals define the different states of sleep.

Rhythm In this book, the word 'rhythm' refers to an internally generated oscillation in some function of a living organism, which is an adaptation to a geophysical cycle. This terminology is used to differentiate internal 'rhythms' from external 'cycles' (see the definition of cycle).

Shift work Any work pattern that requires you to be awake when you would normally be asleep if you were free to choose your own schedule.

Sleep debt Pressure for sleep that builds up over multiple days when sleep is too short or of poor quality. Mounting sleep debt is accompanied by degraded waking function and can only be paid back by sleeping.

Time isolation An experimental protocol where environmental time cues are excluded so that the properties of the internal timekeeping systems of organisms can be investigated.

Window of circadian low (WOCL) Several hours (roughly 3–5 am for most people) when the circadian master clock's drive to wake-promoting centres in the brain is minimal and sleepiness peaks. Around this time, people are at their least functional and most error prone.

References

Chapter One: Living in Earth's Geophysical Cycles

1. New Zealand Department of Conservation. (n.d.). *History of Tiritiri Matangi*. https://www.doc.govt.nz/parks-and-recreation/places-to-go/auckland/places/tiritiri-matangi-scientific-reserve-open-sanctuary/history/
2. Guo, J.-H., Qu, W. M., Chen, S. G., Chen, X. P., Lv, K., Huang, Z. L., & Wu, Y. L. (2014). Keeping the right time in space: Importance of circadian clock and sleep for physiology and performance of astronauts. *Military Medical Research* 1(23). http://www.mmrjournal.org/content/1/1/23
3. Raible, F., Takekata, H., & Tessmar-Raible, K. (2017). An overview of monthly rhythms and clocks. *Frontiers in Neurobiology* 8(189). https://doi.org/10.3389/fneur.2017.00189
4. Chabot, C. C., & Watson, W. H. (2010). Circatidal rhythms of locomotion in the American horseshoe crab *Limulus polyphemus*: Underlying mechanisms and cues that influence them. *Current Zoology* 56(5), 499–517. https://doi.org/10.1093/czoolo/56.5.499
5. Raible et al., (2017). An overview of monthly rhythms and clocks.
6. Rakova, N., Jüttner, K., Dahlmann, A., Schröder, A., Linz, P., Kopp, C., Rauh, M., Goller, U., Beck, L., Agureev, A., Vassilieva, G., Lenkova, L., Johannes, B., Wabel, P., Moissl, U., Vienken, J., Gerzer, R., Eckardt, K.-U., Müller, D. N., ... Titze, J. (2013). Long-term spaceflight simulation reveals infradian rhythmicity in human Na+ balance. *Cell Metabolism* 17(1), 125–131. https://doi.org/10.1016/j.cmet.2012.11.013
7. Raible et al., (2017). An overview of monthly rhythms and clocks.
8. Rahui, G., Nolan, D. R., Bux, D. A., & Schneeberger, A. R. (2019). Is it the moon? Effects of lunar cycle on psychiatric admissions, discharges and length of stay. *Swiss Medical Weekly* 149, w20070. https://doi.org/10.4414/smw.2019.20070
9. Wehr, T. (2018). Bipolar mood cycles associated with lunar entrainment of a circadian rhythm. *Translational Psychiatry* 8(1). https://doi.org/10.1038/s41398-018-0203-x
10. Malpaux, B., Viguie, C., Skinner, D. C., Thiery, J. C., & Chemineau, P. (1997). Control of the annual rhythm of reproduction by melatonin in the ewe. *Brain Research Bulletin* 44(4), 431–438. https://doi.org/10.1016/S0361-9230(97)00223-2
11. Lincoln, G. (2019). A brief history of circannual time. *Journal of Neuroendocrinology* 31(3). https://doi.org/10.1111/jne.12694
12. Ibid.
13. Ibid.
14. Alvarado, S., Mak, T., Liu, S., Storey, K. B., & Szyf, M. (2015). Dynamic changes in global and gene-specific DNA methylation during hibernation in adult thirteen-lined ground squirrels, *Ictidomys tridecemlineatus*. *Journal of Experimental Biology* 218, 1787–1796. https://doi.org/10.1242/jeb.116046
15. West, A., & Wood, S. (2018). Seasonal physiology: Making the future a thing of the past. *Current Opinion in Physiology* 5, 1–8. https://doi.org/10.1016/j.cophys.2018.04.006
16. Cain, S. W., McGlashan, E. M., Vidafar, P., Mustafovska, J., Curran, S. P. N., Wang, X., Mohamed, A., Kalavally, V., & Phillips, A. J. K. (2020). Evening home lighting adversely

impacts the circadian system and sleep. *Scientific Reports 10*(1). https://doi.org/10.1038/s41598-020-75622-4
17. Hazlerigg, H., Blix, A. S., & Stokkan, K.-A. (2017). Waiting for the sun: The circannual programme of reindeer is delayed by the recurrence of rhythmical melatonin secretion after the arctic night. *Journal of Experimental Biology 220*(21), 3869–3872. https://doi.org/10.1242/jeb.163741
18. West & Wood, (2018). Seasonal physiology.
19. Roenneberg, T. (2004). The decline in human seasonality. *Journal of Biological Rhythms 19*, 193–195. https://doi.org/10.1177/0748730404264863
20. Foster, R., & Roenneberg, T. (2008). Human responses to the geophysical daily, annual and lunar cycles. *Current Biology 18*(17), R784–R794. https://doi.org/10.1016/j.cub.2008.07.003
21. Condon, R. G., & Scaglion, R. (1982). The ecology of human birth seasonality. *Human Ecology 10*, 495–511. http://www.jstor.org/stable/4602674
22. Meyer, C., Muto, V., Jaspar, M., Kussé, C., Lambot, E., Chellappa, S. L., Degueldre, C., Balteau, E., Luxen, A., Middleton, B., Archer, S. N., Collette, F., Dijk, D.-J., Phillips, C., Maquet, P., & Vandewalle, G. (2016). Seasonality in human cognitive brain responses. *Proceedings of the National Academy of Sciences USA 113*(11), 3066–3067. https://doi.org/10.1073/pnas.1518129113
23. Wirz-Justice, A. (2018). Seasonality in affective disorders. *General and Comparative Endocrinology 258*, 244–249. https://doi.org/10.1016/j.ygcen.2017.07.010
24. Dopico, X. C., Evangelou, M., Ferreira, R. C., Guo, H., Pekalski, M. L., Smyth, D. J., Cooper, N., Burren, O. S., Fulford, A. J., Hennig, B. J., Prentice, A. M., Ziegler, A.-G., Bonifacio, E., Wallace, C., & Todd, J. A. (2015). Widespread seasonal gene expression reveals annual differences in human immunity and physiology. *Nature Communications 6*, 7000. https://doi.org/10.1038/ncomms8000
25. Wyse, C. A., Morales, C. A. C., Ward, J., Lyall, D., Smith, D. J., Mackay, D., Curtis, A. M., Bailey, M. E. S., Biello, S., Gill, J. M. R., & Pell, J. P. (2018). Population-level seasonality in cardiovascular mortality, blood pressure, BMI and inflammatory cells in UK Biobank. *Annals of Medicine 50*(5), 410–419. https://doi.org/10.1080/07853890.2018.1472389
26. Wirz-Justice, (2018). Seasonality in affective disorders.
27. Stothard, E. R., McHill, A. W., Depner, C. M., Birks, B. R., Moehlman, T. M., Ritchie, H. K., Guzzetti, J. R., Chinoy, E. D., LeBourgeois, M. K., Axelsson, J., & Wright Jr, K. P. (2017). Circadian entrainment to the natural light-dark cycle across seasons and the weekend. *Current Biology 27*(4), 508–513. https://doi.org/10.1016/j.cub.2016.12.041
28. National Aeronautics and Space Administration (2005, 20 March). *Chíchén Itzá*. Image of the day. https://earthobservatory.nasa.gov/images/5349/chichen-itza

Chapter Two: How Our Bodies Keep Time across the Day/Night Cycle

1. Foer, J., & Siffre, M. (2008). Caveman: An interview with Michel Siffre. *Cabinet 30*. https://www.cabinetmagazine.org/issues/30/foer_siffre.php
2. Wever, R. A. (1979). *The circadian system of man: Results of experiments under temporal isolation*. Springer.
3. Gander, P. H., Graeber, R. C., Foushee, H. C., Lauber, J. K., & Connel, L. J. (1994). Crew factors in flight operations II: Psychophysiological responses to short-haul air transport operations. *NASA Technical Memorandum 108856*. NASA Ames Research Centre,

Mountain View, California. https://rosap.ntl.bts.gov/view/dot/12819
4. Weaver, D. R. (1998). The suprachiasmatic nucleus: A 25-year retrospective. *Journal of Biological Rhythms* 13(2), 100–112. https://doi.org/10.1177/074873098128999952
5. Lydic, R., Schoene, W., Czeisler, C., & Moore-Ede, M. (1980). Suprachiasmatic region of the human hypothalamus: Homolog to the primate circadian pacemaker? *Sleep* 2(3), 355–361. https://doi.org/10.1093/sleep/2.3.355
6. Astiz, M., Heyde, I., & Oster, H. (2019). Mechanisms of communication in the mammalian timing system. *International Journal of Molecular Sciences* 20(2), 343. https://doi.org/10.3390/ijms20020343
7. Royal Society of New Zealand Te Apārangi. (2018). *Blue light Aotearoa evidence summary*. https://www.royalsociety.org.nz/assets/Uploads/Blue-light-Aotearoa-evidence-summary.pdf
8. Duffy, J. F., Cain, S. W., Chang, A.-M., & Czeisler, C. A. (2011). Sex difference in the near 24-hour intrinsic period of the human circadian timing system. *Proceedings of the National Academy of Sciences* 108(suppl.3). https://doi.org/10.1073/pnas.1010666108
9. Jagganath, A., Taylor, L., Wakaf, Z., Vasudevan, S. R., & Foster, R. G. (2017). The genetics of circadian rhythms, sleep and health. *Human Molecular Genetics* 26(R2), R128–R138. https://doi.org/10.1093/hmg/ddx240
10. Paine, S. J., Gander, P. H., & Travier, N. (2006). The epidemiology of morningness/eveningness: Influence of age, gender, ethnicity, and socioeconomic factors in adults (30–49 years). *Journal of Biological Rhythms* 21(1), 68–76. https://doi.org/10.1177/0748730405283154
11. Sleep Health Foundation. (2021). *Melatonin and sleep*. www.sleepfoundation.org/melatonin
12. Roenneberg, T., Winnebeck, E., & Klerman, E. (2019). Daylight saving and artificial time zones: A battle between biological and social times. *Frontiers in Physiology* 10. https://doi.org/10.3389/fphys.2019.00944
13. Ibid.
14. Ibid.
15. Paine, S. J., & Gander, P. H. (2016). Explaining ethnic inequities in sleep duration: A cross-sectional survey of Māori and non-Māori adults in New Zealand. *Sleep Health* 2(2), 109–115. https://doi.org/10.1016/j.sleh.2016.01.005
16. Parsons, M. J., Moffitt, T. E., Gregory, A. M., Goldman-Mellor, S., Nolan, P. M., Poulton, R., & Caspi, A. (2015). Social jetlag, obesity and metabolic disorder: Investigation in a cohort study. *International Journal of Obesity* 39(5), 842–848. https://doi.org/10.1038/ijo.2014.201

Chapter Three: Sleep – How We Know What We Know

1. Aserinsky, E., & Kleitman, N. (1953). Regularly occurring periods of eye motility, and concomitant phenomena, during sleep. *Science* 118(3062), 273–274. https://doi.org/10.1126/science.118.3062.273
2. Dement, W. C., Kushida, C. A., & Chang, J. (2004). History of sleep deprivation. In Kushida, C. (Ed.). *Sleep deprivation: Basic science, physiology, and behavior* (31–46). Taylor & Francis.
3. Prabhavananda, S., & Manchester, F. (1983). *The Upanishads: Breath of the eternal* (English translation). Vedanta Press.
4. Jouvet, M. (1996). *Le Songe de Cro-magnon*. Science et Avenir Hors-Série Le Rêve (my

translation). http://sommeil.univ-lyon1.fr/articles/savenir/cromagnon/cromagnon.php
5. Ibid.
6. Hobson, J. A. (2003). *Dreaming: A very short introduction*. Oxford University Press.
7. Ibid.
8. Cartwright, R. D. (2010). *The twenty-four-hour mind: The role of sleep and dreaming in our emotional lives*. Oxford University Press.
9. Stickgold, R. (2016). Psychobiology and dreaming: Introduction. In Kryger, M., Roth, T., & Dement, W. C. (Eds.). *Principles and practice of sleep medicine* (6th ed.) (506–508). Elsevier.
10. Ekirch, R. A. (2001). Sleep we have lost: Pre-industrial slumber in the British Isles. *American Historical Review* 106(2), 343–386. https://doi.org/10.1086/ahr/106.2.343; Ekirch, R. A. (2005). Sleep we have lost: Rhythms and revelations. In *At day's close: Night in times past* (300–323). Norton.
11. Wehr, T. A. (1992). In short photoperiods, human sleep is biphasic. *Journal of Sleep Research* 1(2), 103–117. https://doi.org/10.1111/j.1365-2869.1992.tb00019.x
12. Yetish, G., Kaplan, H., Gurven, M., Wood, B., Pontzer, H., Manger, P. R., Wilson, C., McGregor, R., & Siegel, J. M. (2015). Natural sleep and its seasonal variations in three pre-industrial societies. *Current Biology* 25(21), 2862–2868. https://doi.org/10.1016/j.cub.2015.09.046
13. Ekirch, (2001). Sleep we have lost.
14. Ibid.
15. Wehr, T. A. (2001). Photoperiodism in humans and other primates: Evidence and implications. *Journal of Biological Rhythms* 16(4), 348–364. https://doi.org/10.1177/074873001129002060
16. Roth, T., Berglund, P., Shahly, V., Shillington, A. C., Stephenson, J. J., & Kessler, R. C. (2013). Middle-of-the-night hypnotic use in a large national health plan. *Clinical Sleep Medicine* 9(7), 661–668. https://doi.org/10.5664/jcsm.2832
17. Yetish et al., (2015). Natural sleep and its seasonal variations.

Chapter Four: Sleep – Why We Need It

1. Fultz, N. E., Bonmassar, G., Setsompop, K., Stickgold, R. A., Rosen, B. R., Polimeni, J. R., & Lewis, L. D. (2019). Coupled electrophysiological, hemodynamic, and cerebrospinal fluid oscillations in human sleep. *Science* 336(6465), 628–631. https://doi.org/10.1126/science.aax5440
2. Murkar, A., & De Konink, J. (2018). Consolidative mechanisms of emotional processing in REM sleep and PTSD. *Sleep Medicine Reviews* 41, 173–184. https://doi.org/10.1016/j.smrv.2018.03.001
3. Wei, L., Lei, M., Yang, G., & Wen-Biao, G. (2017). REM sleep selectively prunes and maintains new synapses in development and learning. *Nature Neuroscience* 20, 427–437. https://doi.org/10.1038/nn.4479
4. Banks, S., Dorrian, J., Basner, M., & Dinges, D. F. (2017). Sleep deprivation. In Kryger et al., *Principles and practice of sleep medicine* (49–55).
5. Royal Society for Public Health. (n.d.). *Waking up to the health benefits of sleep*. https://www.rsph.org.uk/resourceLibrary/waking-up-to-the-health-benefits-of-sleep.html
6. Lange, T., Dimitrov, S., Bollinger, T., Diekelmann, S., & Born J. (2011). Sleep after vaccination boosts immunological memory. *The Journal of Immunology* 187(1), 283–290. https://doi.org/10.4049/jimmunol.1100015

7. Buysse, D. J. (2014). Sleep health: Can we define it? Does it matter? *Sleep* 37(1), 9–17. https://doi.org/10.5665/sleep.3298
8. Hirshkowitz, M., Whiton, K., Albert, S. M., Alessi, C., Bruni, O., DonCarlos, L., Hazen, N., Herman, J., Hillard, P. J. A., Katz, E. S., Kheirandish-Gozal, L., Neubauer, D. N., O'Donnell, A. E., Ohayon, M., Peever, J., Rawding, R., Sachdeva, R. C., Setters, B., Vitiello, M. V., & Ware, J. C. (2015). National Sleep Foundation's updated sleep duration recommendations: Final report. *Sleep Health* 1(4), 233–243. https://doi.org/10.1016/j.sleh.2015.10.004
9. Royal Society for Public Health. (n.d.). *Waking up to the health benefits of sleep.*
10. Ibid.
11. Lee, C. J. (2016). Sleep: A human rights issue. *Sleep Health* 2(1), 6–7. https://doi.org/10.1016/j.sleh.2015.12.007
12. Hillman, D., Mitchell, S., Streatfeild, J., Burns, C., Bruck, D., & Pezzullo, L. (2018). The economic cost of inadequate sleep. *Sleep* 41(8), 1–13. https://doi.org/10.1093/sleep/zsy083
13. Dement, W. C. (2000). *The promise of sleep*. Dell.

Chapter Five: Circadian Biology versus 24/7 Living

1. Gander, P. H., O'Keeffe, K., Santos-Fernandez, E., Huntington, A., Walker, L., & Willis, J. (2020). Development and evaluation of a matrix for assessing fatigue-related risk, derived from a national survey of nurses' work patterns. *International Journal of Nursing Studies* 112(103573). https://doi.org/10.1016/j.ijnurstu.2020.103573
2. Gander, P. H., Gregory, K. B., Connell, L. J., Miller, D. L., Graeber, R. C., & Rosekind, M. R. (1996). Crew factors in flight operations VII: Psychophysiological responses to overnight cargo operations. *NASA Technical Memorandum 110380*. https://ntrs.nasa.gov/archive/nasa/casi.ntrs.nasa.gov/19960016648.pdf
3. Folkhard, S. (2008). Do permanent night workers show circadian adjustment? A review based on the endogenous melatonin rhythm. *Chronobiology International* 25(2), 215–224. https://doi.org/10.1080/07420520802106835
4. Klein, K. E., Herrmann, R., Kuklinski, P., & Wegmann, H. M. (1977). Circadian performance rhythms: Experimental studies in air operations. In Mackie, R. R. (Ed.). *Vigilance*. NATO Conference Series, vol. 3. Springer.
5. Gander, P. H., Gregory, K. B., Graeber, R. C., & Connell, L. J. (1991). Crew factors in flight operations VIII: Factors influencing sleep timing and subjective sleep quality in commercial long-haul flight crews. *NASA Technical Memorandum 103852*. https://ntrs.nasa.gov/citations/19920010735
6. Gander, P., Mulrine, H. M., van den Berg, M. J., Wu, L., Smith, A., Signal, L., & Mangie, J. (2016). Does the circadian clock drift when pilots fly multiple transpacific flights with 1- to 2-day layovers? *Chronobiology International* 33(8), 982–994. https://doi.org/10.1080/07420528.2016.1189430
7. IARC Monographs, vol. 124 group. (2019). Carcinogenicity of night shift work. *Lancet Oncology* 20(8), 1058–1059. https://doi.org/10.1016/S1470-2045(19)30455-3
8. Fransen, M., Wilsmore, B., Winstanley, J., Woodward, M., Grunstein, R., Ameratunga, S., & Norton, R. (2006). Shift work and work injury in the New Zealand Blood Donors' Health Study. *Occupational and Environmental Medicine* 63(5), 352–358. https://doi.org/10.1136/oem.2005.024398
9. Marquié, J.-C., Tucker, P., Folkhard, S., Gentil, C., & Ansiau, D. (2015). Chronic effects

of shift work on cognition: Findings from the VISAT longitudinal study. *Occupational and Environmental Medicine* 72(4). https://doi.org/10.1136/oemed-2013-101993
10. Gander, P., O'Keeffe, K., Santos-Fernandez, E., Huntington, A., Walker, L., & Willis, J. (2019). Fatigue and nurses' work patterns: An online questionnaire survey. *International Journal of Nursing Studies* 98, 67–74. https://doi.org/10.1016/j.ijnurstu.2019.06.011
11. Dejohn, C. (2020). President's Page: Focus on the Sleep/Wake Research Centre, Massey University, New Zealand. *Aerospace Medicine and Human Performance* 91(9), 687–688. https://doi.org/10.3357/AMHP.9109PP.2020
12. International Air Transport Association (IATA), International Civil Aviation Organisation & International Federation of Airline Pilots' Associations. (2015). *Fatigue management guide for airline operators* (2nd ed.). IATA.
13. Safer Nursing 24/7 Project. (2019). National Code of Practice for Managing Nurses' Fatigue and Shift Work in District Health Board Hospitals. https://www.safernursing24-7.co.nz
14. Nurses' Health Study. (n.d.). Nurses' Health Study II. www.nurseshealthstudy.org
15. Jørgensen, J. T., Karlsen, S., Stayner, L., Hansen, J., & Andersen, Z. J. (2017). Shift work and overall cause-specific mortality in the Danish nurse cohort. *Scandinavian Journal of Work Environment and Health* 43(2), 117–126. https://doi.org/10.5271/sjweh.3612
16. McHill, A. W., & Wright, K. P. (2017). Role of sleep and circadian disruption on energy expenditure and in metabolic predisposition to human obesity and metabolic disease. *Obesity Reviews* 18(suppl.1), 15–24. https://doi.org/10.1111/obr.12503
17. Zitting K.-M., Vetrivelan, R., Yuan, R. K., Duffy, J. F., Saper, C. B., & Czeisler, C. A. (2022). Chronic circadian disruption on a high-fat diet impairs glucose tolerance. *Metabolism* 130(155158). https://doi.org/10.1016/j.metabol.2022.155158
18. El-Sokkary, G. H., Ismail, I. A., & Saber, S. H. (2018). Melatonin inhibits breast cancer cell invasion through modulating DJ-1/KLF17/ID-1 signaling pathway. *Journal of Cellular Biochemistry* 120(3), 3945–3957. https://doi.org/10.1002/jcb.27678
19. As of May 2020, the countries that had ratified C171 were Albania, Belgium, Brazil, Chzechia, Côte d'Ivoire, Cyprus, Dominican Republic, Lao People's Democratic Republic, Lithuania, Luxembourg, Madagascar, Montenegro, North Macedonia, Portugal, Slovakia, Slovenia, and Uruguay.
20. IARC Monographs, vol. 124 group. (2019). Carcinogenicity of night shift work.
21. Rogers, W. P., Acheson, D. C., Armstrong, N. A., Covert, E. E., Feynman, R. P., Hotz, R. B., Kutyna, D. J., Ride, S. K., Rummel, R. W., Sutter, J. F., Walker, A. B. C., Wheelon, A. D., & Yeager, C. E. (1986). *Report of the Presidential Commission on the Space Shuttle Challenger accident*. Appendix G: Human factors analysis. https://history.nasa.gov/rogersrep/genindex.htm
22. Ibid.

Chapter Six: Our Home Planet

1. Gaston, K. J., Visser, M. E., & Hölker, F. (2015). The biological impacts of artificial light at night: The research challenge. *Philosophical Transactions of the Royal Society B: Biological Sciences* 370(1667). https://doi.org/10.1098/rstb.2014.0133
2. Royal Society of New Zealand Te Apārangi. (2018). *Blue light Aotearoa evidence summary*. https://www.royalsociety.org.nz/assets/Uploads/Blue-light-Aotearoa-evidence-summary.pdf
3. Owens, A. C. S., & Lewis, S. M. (2018). The impact of artificial light at night on

nocturnal insects: A review and synthesis. *Ecology and Evolution* 8(22), 11337–11358. https://doi.org/10.1002/ece3.4557

4. Spoelstra, K., van Grunsven, R. H. A., Donners, M., Gienapp, P., Huigens, M. E., Slaterus, R., Berendse, F., Visser, M. E., & Veenendaal, E. (2015). Experimental illumination of natural habitat – an experimental set-up to assess the direct and indirect ecological consequences of artificial light of different spectral composition. *Philosophical Transactions of the Royal Society B: Biological Sciences* 370(1167). https://doi.org/10.1098/rstb.2014.0129
5. ffrench-Constant, R. H., Somers-Yeates, R., Bennie, J., Economou, T., Hodgson, D., Spalding, A., & McGregor, P. K. (2016). Light pollution is associated with earlier tree budburst across the United Kingdom. *Proceedings of the Royal Society B: Biological Sciences* 283(1833). https://doi.org/10.1098/rspb.2016.0813
6. Gessler, A., Bugmann, H., Bigler, C., Edwards, P., della Guistina, C., Kueffer, C., Roy, J., & Resco de Dios, V. (2017). Light as a source of information in ecosystems. In Sanders, S., & Oberst, J. (Eds.). *Changing perspectives on daylight: Science, technology and culture* (9–15). Science/AAAS.
7. Spoelstra et al., (2015). Experimental illumination of natural habitat.
8. Dominoni, D. M., Jensen, J. K., de Jong, M., Visser, M. E., & Spoelstra, K. (2020). Artificial light at night, in interaction with spring temperature, modulates timing of reproduction in a passerine bird. *Ecological Applications* 30(3). https://doi.org/10.1002/eap.2062
9. Gaston et al., (2015). The biological impacts of artificial light at night.
10. Davies, T. W., Duffy, J. P., Bennie, J., & Gaston, K. J. (2014). The nature, extent and ecological implications of marine light pollution. *Frontiers in Ecology and the Environment* 12(6), 347–355. https://doi.org/10.1890/130281
11. Witherington, B. E., & Martin, R. E. (2003). *Understanding, assessing, and resolving light-pollution problems on sea turtle nesting beaches* (3rd ed., rev.) (73). Florida Marine Research Institute, Technical Report TR-2.
12. Knoop, M., Stefani, O., Bueno, B., Matusiak, B., Hobday, R., Wirz-Justice, A., Martiny, K., Kantermann, T., Aarts, M. P. J., Zemmouri, N., Appelt, S., & Norton, B. (2020). Daylight: What makes the difference? *Lighting Research and Technology* 52(3), 423–442. https://doi.org/10.1177/1477153519869758
13. Royal Society of New Zealand Te Apārangi. (2018). *Blue light Aotearoa evidence summary*.
14. Knoop et al., (2020). Daylight: What makes the difference?
15. Samel, A., & Gander, P. (1991). Light as a chronobiologic countermeasure for long-duration space operations. *NASA Technical Memorandum* 103874, 1–70. https://ntrs.nasa.gov/api/citations/19920021923/downloads/19920021923.pdf
16. Basner, M., Dinges, D. F., Mollicone, D., Ecker, A., Jones, C. W., Hyder, E. C., Di Antonio, A., Savelev, I., Kan, K., Goel, N., Morukov, B. V., & Sutton, J. P. (2013). Mars 520-d simulation reveals protracted crew protracted crew hypokinesis and alterations of sleep duration and timing. *Proceedings of the National Academy of Sciences USA* 110(7), 2635–2640. https://doi.org/10.1073/pnas.1212646110
17. Flynn-Evans, E., Gregory, K., Arsintescu, L., & Whitmire, A. (2016). Risk of performance decrements and adverse health outcomes resulting from sleep loss, circadian desynchronization, and work overload. *NASA Technical Report* JSC-CN-35774. https://ntrs.nasa.gov/citations/20160003864

Index

Entries in **bold** refer to image captions

Philippa Gander is a professor emeritus at Massey University and an internationally recognised scholar of sleep and circadian rhythms. After a PhD at the University of Auckland and research at Harvard University Medical School and the NASA Ames Research Center, Gander was the inaugural director of the Sleep/Wake Research Centre at Massey University. In 2009 she was elected a Fellow of the Royal Society of New Zealand and in 2017 she was appointed an Officer of the New Zealand Order of Merit, for services to the study of sleep and fatigue.